PRAISE FOR *HIGH PLAINS HOMESTEAD*

"A well written and researched book that depicts the challenges of farming in the often foreboding environment of Central Kansas. Charts and photos explained topics ranging from mechanization and economic cost to commodity prices and a decline in the rural population. It gave a very personal insight of the extended family of the author covering the period of the late 1870s to the present time. It was a thought provoking book with ongoing challenges in each chapter, ending with the tragedy of an immense fire in Northern Russell County as well as three other counties.

I personally read the book in two days and found myself immersed in all areas/topics covered in the book, partly because it reflected on much of my early life in Russell County."

—*Galen Niedenthal, past President Russell County,*
Kansas Farm Bureau

"The Crawford family, from their 1879 grassland acreage to present, have witnessed much evolution in their lives—from horse power to steam power to petroleum power to solar and wind power. The next phase, robotic and artificial intelligence leave a lot to anticipate.

The book, well-illustrated and researched, depicts the character, ambition, and the love of the land that the Crawford family had to transform their small part of the central Kansas grassland in 1879 into the successful farm entity it is today."

—*Delmar Hampl, retired farmer and lifelong acquaintance*

"A remarkable book that combines family history in the high plains of Kansas with the evolving nature of agriculture. Crawford portrays the hard work and determination of men and women in his family, along with their neighbors, who settled the land and formed communities. Four generations of Crawfords farmed this land, and it was designated A Century Farm. This book makes a significant contribution to the recorded history of Kansas."

—Ruth Johnson Smiley, PhD, member of First Families of Tennessee and co-editor of Tennessee Women of Vision and Courage

"Maps, charts, and many photographs woven into the text help make this story clear, easily readable, and a valuable addition to the history of the central Kansas prairie."

—Kevin Hampl, present day farmer of this Kansas land

"Kent Crawford has written about the birth and transformation of his family farm in Russell County through four generations. The book has factual and technical information but enough of his personal story to make it interesting for the reader. It will appeal especially to those having lived or living in this area or those with similar stories in their family histories. The book brings into question the future of the American family farm."

—Aldean Banker, former board member of Russell County Historical Society, Resident of Russell County for more than 65 years

HIGH PLAINS
HOMESTEAD

Also by R. Kent Crawford

Ruts, Guts, and a Model T Truck

HIGH PLAINS
HOMESTEAD

*Evolution of a
Century Farm*

R. KENT CRAWFORD

First Edition
ISBN: 979-8-9875793-0-5
Library of Congress Control Number: Application submitted

Post Rock Press
P.O. Box 24314
Knoxville, TN 37933

We come and go, but the land is always here. And the people who love it and understand it are the people who own it—for a little while.

—Willa Cather, *O Pioneers!*

CONTENTS

Century Farm Family

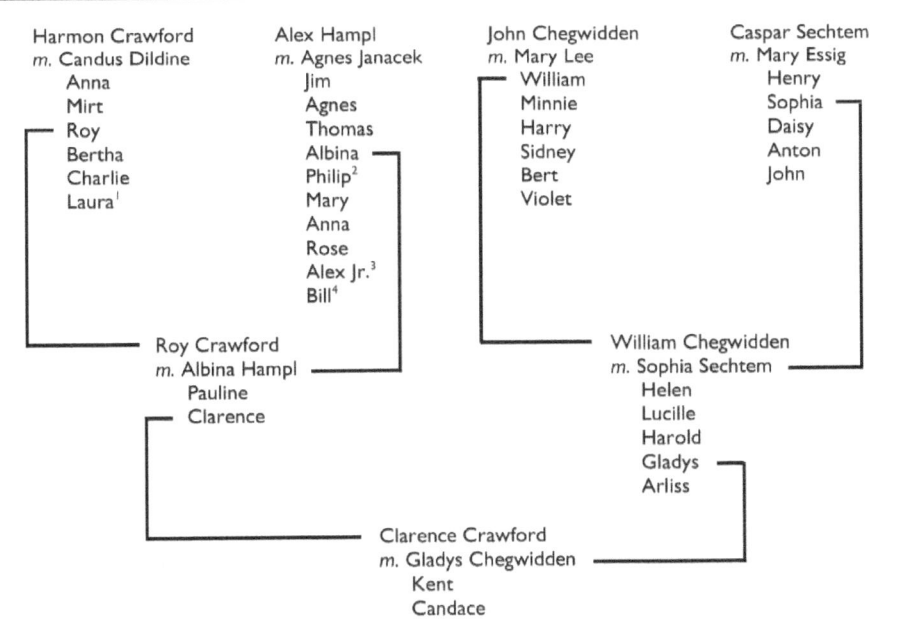

Harmon Crawford	Alex Hampl	John Chegwidden	Caspar Sechtem
m. Candus Dildine	m. Agnes Janacek	m. Mary Lee	m. Mary Essig
Anna	Jim	William	Henry
Mirt	Agnes	Minnie	Sophia
Roy	Thomas	Harry	Daisy
Bertha	Albina	Sidney	Anton
Charlie	Philip[2]	Bert	John
Laura[1]	Mary	Violet	
	Anna		
	Rose		
	Alex Jr.[3]		
	Bill[4]		

Roy Crawford
m. Albina Hampl
Pauline
Clarence

William Chegwidden
m. Sophia Sechtem
Helen
Lucille
Harold
Gladys
Arliss

Clarence Crawford
m. Gladys Chegwidden
Kent
Candace

1. Laura (Crawford) Luder provided information about the early days of the Crawfords.

2. Philip Hampl and Mathilda (Bachmann) Hampl were the parents of Delmar Hampl. Delmar farmed the Crawford land for a number of years, and also provided historical information about the Crawfords and Hampls. Delmar's son Kevin Hampl currently farms the Crawford land.

3. Alex Hampl Jr. and Sarah (Griffin) Hampl were the parents of Eldon Hampl, who farmed Pauline's land for several years. Eldon also supplied pictures of the early days of the town of Luray.

4. Bill Hampl provided information about the early days of the Hampl family. Bill was married to Helen Chegwidden.

A New Beginning

PROMISE OF THE FRONTIER

A Brief Interlude

Three minutes to go!

What could be better than a crisp afternoon in mid-November, perfect football weather, and the Luray team with a comfortable lead over our archrival Lucas? Most of the townsfolk of Luray and a good many of the country folk as well lined the sidelines to cheer us on. We were anxious to finish the game so we could begin to celebrate, but the clock dragged on excruciatingly as we ran play after play, marching down the field once again as time ran out. Finally, the referee blew his whistle and the game was over. Luray had defeated Lucas 36-7 to cap off a perfect 9-0 football season. With a roar from the crowd and a honking of horns, for a short while, all was right with the world.

Football was just a high school sport, but the whole town of Luray and the surrounding countryside considered this game and this season to be wins for them as well. Football had united the community, giving it something positive to celebrate and talk about rather than the usual worries about whether the drought would continue, if the wheat would get a good start that fall, and how bad the winter would be.

The year was 1957, and I was a junior at Luray High School and a member of that winning football team. I lived on a farm southwest of the town of Luray with my sister Candace and my parents Clarence and Gladys. For me at that time, being on that victorious football team was just about the most important thing in the world. However, over time I came to realize that our farm represented a far greater series of victories, extending over more than a century, spanning four generations of the Crawford family, and culminating with the Kansas Farm Bureau designating our farm a Century Farm. This history of that family farm and the associated community begins well before that memorable football season, extends well beyond that season, and explores the major transitions that occurred along the way.

The Quest for Land

Our story begins when John Crawford, the first of my line of Crawfords in the New World, arrived in America in about 1735, having made the trip from Northern Ireland as part of the wave of Scotch-Irish immigration. In much of Europe in the early 1700s, the aristocracy owned most property. Ownership of land was the source of wealth and power, but tenant farmers, peasants who had neither wealth nor power and struggled to eke out a living under the harsh conditions imposed by the property owners, carried out the actual farming of the land. The Scotch-Irish, originally Scots from the borderlands between Scotland and England, understandably had a desire to escape the frequent fighting along that border. English owners of undeveloped land in Northern Ireland offered low rental fees that enticed some of these Scots to settle on, improve, and farm that land. Generations later, after these immigrants had turned that undeveloped land into productive farmland, the property owners dramatically increased the rents they were charging, driving many of these Scotch-Irish to immigrate once again, this time to America with its open frontier where even peasants could own land.

By the time John Crawford set foot in America, the earlier arrivals from Europe had claimed nearly all the coastal lowlands, and John had to head for less expensive land on the frontier further inland. He settled with his wife Elizabeth in Mount Bethel Township where a Scotch-Irish community was forming on the Pennsylvania side of the Delaware River about 20 miles north of present-day Easton. There John became a successful farmer, acquiring farmland and growing his farm to nearly 200 acres by 1775.

Edmond Crawford, the second in my line of Crawfords in America, was a younger son of John and Elizabeth. As a younger son, he would inherit little or none of John's farmland, and so had to head elsewhere to find his opportunities. By the time Edmond reached manhood, Mount Bethel was beginning to feel crowded, so in 1774 Edmond joined a group of Scotch-Irish who were heading west to the next new frontier, where they could obtain land for little or no monetary cost—one just had to be willing to work hard to turn it into a farm.

Parts of Pennsylvania had been a battleground during the French and Indian War, fought between 1754 and 1763. That war ended with the 1763 Treaty of Paris, and subsequently, King George III of England issued the Royal Proclamation of 1763 reserving all lands west of the Appalachian Mountains to their Native American inhabitants. This proclamation prohibited westward-bound Americans and others from settling in that reserved region, thus defining the Appalachians as the western edge of the colonial frontier. Edmond's group settled at the eastern edge of the Appalachians in an area along the Susquehanna River near present-day Bloomsburg, Pennsylvania. That was on the western frontier, as far west as they could go under the constraints of the Royal Proclamation.

Despite occasional raids by the Native American tribes located along this frontier area, Edmond became a prosperous farmer and raised a family there. Edmond and his wife Mary (Roseberry) had four sons, the first of whom was Joseph born in 1778. Each of those sons lived on and farmed a portion of Edmond's lands; Joseph inherited his portion upon Edmond's death. Joseph Crawford continued to farm in this area, and Joseph II, one of his sons, remained in this area and continued to farm part of his father's land there as well.

The Advancing Frontier

The conclusion of the Revolutionary War opened new lands to settlement. The 1783 Treaty of Paris marking the end of the Revolutionary War gave the United States all the land the English had claimed in the area east of the Mississippi River, north of Florida, and south of Canada. The Congress of the Confederation passed the Northwest Ordinance in 1787, creating the Northwest Territory from the portion of this land that was west of Pennsylvania and northwest of the Ohio River. This Territory included more than 260,000 square miles; a vast wilderness inhabited by about 45,000 Native Americans and fewer than 4,000 traders (mostly French). It comprised about a third of the land area of the United States at that time. The United States nominally "owned" this territory, and it mostly ignored any ownership rights of the Native Americans.

The government set out to populate this land, but before it could open the land to private ownership, it had to survey the territory to provide a basis for identification of ownership. The Land Ordinance of 1785 created a rectangular-grid system, the Public Land Survey System, to survey this new land. In this system, two controlling lines are the first surveyed, an east-west "baseline" and a north-south "principal meridian." Subsequent work relates to these lines, dividing the land into approximately square "survey townships" roughly six miles on each side. A survey township coordinate number (counted from the baseline) and a range coordinate number (counted from the principal meridian) designate the location of each township (figure 1). The surveyors then divide each survey township into 36 square "sections" of approximately one mile on each side, with each section roughly 640 acres in area. Figure 1 also shows the section numbering within the survey township. In this system, the township, range, and section numbers uniquely describe every section. (Note: The townships discussed here are "survey townships" and are distinct from "civil townships," the latter being a local governing entity subordinate to a county.)

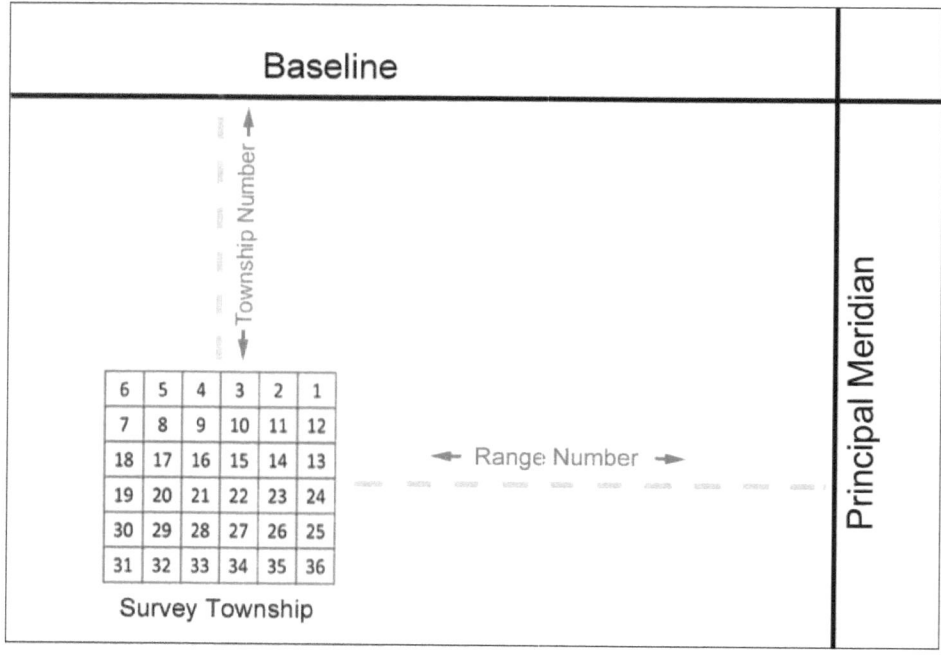

1. Location and section numbering for a survey township.

The frontier moved westward as the survey proceeded, permitting settlement of additional parts of the Northwest Territory. This territory eventually became the states of Ohio (admitted in 1803), Indiana (1816), Illinois (1818), Michigan (1837), Wisconsin (1848), and part of what would later become Minnesota. A few of the Crawford family descendants decided to follow the frontier to find new opportunity in the Northwest Territory during the first half of the nineteenth century, but most of the extended Crawford families chose to remain in their comfortable homes in what had now become Columbia County, Pennsylvania.

In the 1803 Louisiana Purchase, the United States purchased the territory of Louisiana from France. This territory included nearly 530 million acres (828,000 square miles) extending from the Mississippi River to the Rocky Mountains. The United States paid France 15 million dollars for the Louisiana Purchase or about three cents per acre. However, the Louisiana Purchase was not quite the bargain it seemed since France had actual control of only a small part of this territory, and Native American tribes controlled the rest. What the United States bought with this purchase was the exclusive right to deal with the Native Americans to obtain their land by purchase or conquest, without interference from other colonial powers. The initial survey of the Louisiana Purchase lands covered what later became Arkansas, Missouri, Iowa, North Dakota, South Dakota, and Minnesota, and as settlers arrived, the frontier continued to shift westward across these newly surveyed lands.

All this happened before my great-grandparents were born. My great-grandfather, Harmon Labour Crawford, was born in Columbia County, Pennsylvania, in 1849 as one of the sons of Joseph Crawford II, making him a member of the fifth generation of our line of Crawfords in America.

Along the way, marriages had mixed other backgrounds with the original Scotch-Irish of the Crawfords—Harmon was an ethnic mix of Scotch-Irish, English, German, French, Belgian, and Dutch. Some of his ancestors had arrived in America as early as the mid-1600s.

My great-grandmother, Ruth Candus Dildine, was born in 1848 and was one of the seventh generation of Dildines to live in America. Her immediate family had also made their way to Columbia County. Candus's Dildine ancestors came to America in 1710, and some of the families into which they married had arrived even earlier, several as early as the mid-1600s.

The earliest Dildines were German, but by Candus's generation, the addition of other ethnicities in the American melting pot made her a genetic and cultural mixture of German, French, Dutch, English, and Norwegian. Harmon Labour Crawford married Ruth Candus Dildine in Orangeville Pennsylvania in 1870. Harmon and Candus would ultimately make their way to the frontier on the High Plains of Western Kansas, where they would establish our family farm.

Kansas Becomes the New Frontier

In 1854, Congress passed the Kansas-Nebraska Act creating both the Kansas Territory and the Nebraska Territory from a part of the Louisiana Purchase, and established the office of Surveyor General for Kansas and Nebraska to survey these territories. The Kansas Territory extended from the Missouri border on the east to the summit of the Rocky Mountains on the west, and from the thirty-seventh parallel north to the fortieth parallel north.

Within days of the passage of the Kansas-Nebraska Act, the race to settle Kansas was on between settlers from Missouri, which was a slave state, and settlers from the Midwest and Eastern states, which were Free States. This race led to violence on both sides, resulting in the label "Bleeding Kansas."

The Free State settlers ultimately outvoted the pro-slavery settlers, and the eastern portion of the Kansas Territory entered the Union as the Free State of Kansas in 1861. The state of Kansas spanned 400 miles from east to west and 200 miles from north to south. The state did not include the western end of the Kansas Territory, which later became part of the state of Colorado.

The fortieth parallel marked the border between the Kansas and Nebraska Territories and became the baseline for the survey of these territories. The Sixth Principal Meridian for survey purposes, established at a point 108 miles west of the Missouri River as measured along the Kansas-Nebraska border, served as the second controlling survey line. Meridian Avenue in Wichita, Kansas lies along a portion of this Sixth Principal Meridian, and US

Highway 81 is a route that roughly follows this meridian as it crosses the state from north to south. This meridian also marks an unofficial nebulous boundary between Eastern Kansas (about one third of the state) and Western Kansas (nearly two thirds of the state).

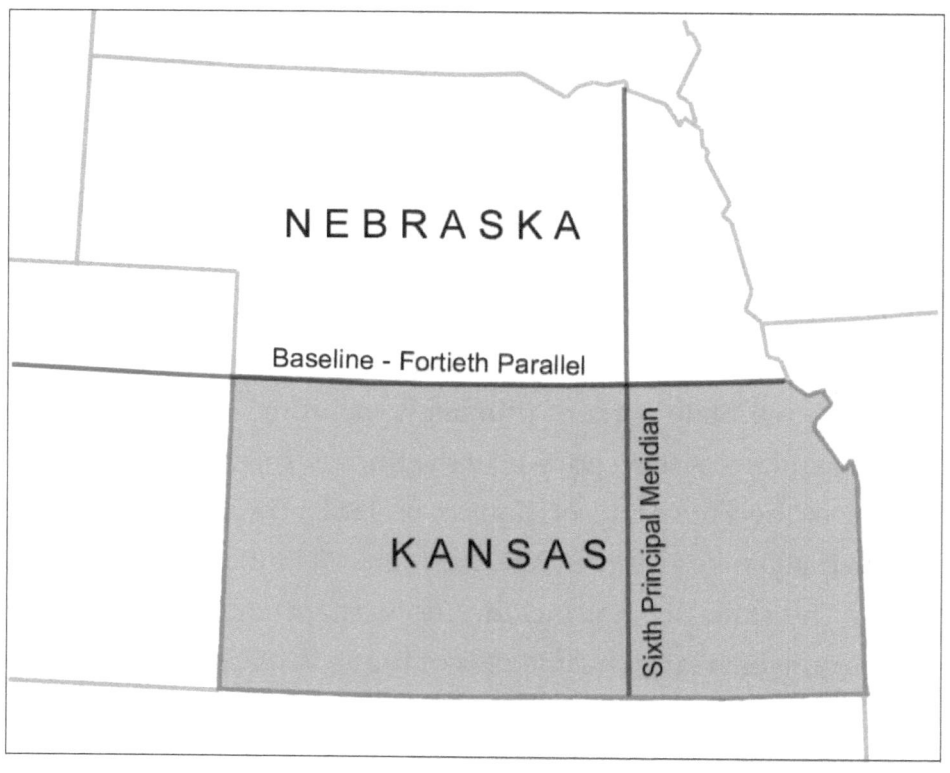

2. Sixth Principal Meridian.

Eastern Kansas is a land of hills, numerous rivers and forests, and plentiful rainfall, like much of the eastern part of the United States. Western Kansas is quite different; the climate grows harsher and the annual rainfall amounts diminish steadily as one proceeds westward from the Sixth Principal Meridian. The average elevation increases at the rate of about eight feet per mile, from about 600 feet at the Missouri River to well over 3,000 feet at the Colorado border.

The land in Western Kansas, part of the High Plains, was primarily short-grass prairie mostly devoid of trees, earning the sobriquet "The Great American Desert." The surveyors had to contend with the extremes of Kansas weather, huge herds of buffalo, and occasional conflicts with the Native American inhabitants, so they required 21 years to survey all of Kansas, with the surveying continuing well beyond the time at which Kansas became a state.

The Homestead Act of 1862, following shortly after Kansas became a state, played a major role in the settlement of Kansas. The Free Soil Party, later merged into the New Republican Party, demanded that the new lands to the west be open to independent small farmers. Southern Democrats continually defeated any such proposal in Congress, since creating many small independent farms would likely prevent the establishment of large plantations on these new lands, and hence would prevent these lands from becoming slave states.

However, once the South seceded from the Union, its members left Congress and the remaining Congressional members easily passed the Homestead Act, which President Lincoln signed in 1862. The Act granted 160 acres of public land in the west to any US citizen willing to settle on and farm the land. This law required a three-step procedure: apply for an "entry permit" authorizing initial settlement on the land, improve the land (live on the designated land, build a home, make improvements, and farm it for a minimum of five years), and file for the patent (deed). Any citizen who had never taken up arms against the US government and was head of a household or was at least 21 years old could apply. The filing fee was $18. After the end of the Civil War, Union soldiers were able to deduct the time they had served from the residency requirement on the land, streamlining the homesteading process for them.

The construction of railroads across Kansas began shortly after Kansas became a state, and these railroads facilitated the movement of settlers into the state. The US Government authorized the original Union Pacific Rail Road, a portion of the first transcontinental railroad, in 1862. Construction began in 1865, starting from Council Bluffs, Iowa along the Missouri River. It made its way westward across Nebraska, passing through Omaha, Fremont, Grand Island, Kearney, and North Platte before entering Wyoming at Pine Bluffs. That railroad never entered Kansas. However, in 1863 the Pacific Railway Act authorized the Union Pacific Eastern Division Railroad (not affiliated with the Union Pacific Railroad) to create a southern branch of the transcontinental railroad across Kansas, in parallel with the Union Pacific route across Nebraska.

The Union Pacific Eastern Division began construction starting from Kansas City in 1863, and by 1864, the line was operational to Lawrence, Kansas. The rest of the route across Kansas proceeded rapidly, reaching Junction City in 1866, Salina in 1867, and Denver,

Colorado, in 1870. The Union Pacific Eastern Division, renamed the Kansas Pacific Railway in 1869, consolidated with the Union Pacific in 1880. With the completion of the Kansas Pacific Railroad through Kansas, the government granted the railroad alternate sections of land in a band extending in a checkerboard pattern 20 miles on each side of the track. This amounted to 20 sections per mile of track laid. The railroad sold parcels of this "railroad land" to settlers who could afford it, and this was an alternative to homesteading.

The population of Kansas, a little over 100,000 in 1860, had undergone a three-fold increase to 364,000 inhabitants by 1870, most of whom were in the eastern part of the state. However, the completion of the railroad across Kansas, coupled with the availability of free homestead land and inexpensive land sold by the railroads, and the end of the Civil War, led to a dramatic increase in the influx of settlers between 1870 and 1890. The population of Kansas reached nearly one million in 1880 and exceeded 1.4 million in 1890. This burst of settlement was predominantly over by 1890, with only slow population growth occurring after that time. Naturally, the earlier settlers got land in the best locations for farming—bottom land along the rivers and land in the eastern part of the state where rainfall was more plentiful. By the late 1870s most of the remaining available land was in the western portion of the state in the vast grasslands of the mostly treeless prairie.

These western lands noted for their dry and windy climate and for their frequent violent storms, fall within the zone of Prevailing

Westerlies, where the wind would normally be blowing from southwest to northeast. However, the Rocky Mountains lying just to the west of Kansas interact with these westerly winds and profoundly affect the climate of Kansas and the rest of the Great Plains. As the westerly wind passes over the tops of the mountains it descends into the space available below it, warming and gaining speed as it rushes down the eastern slope of the mountains to become a strong west wind as it reaches the plains. These hot dry winds, averaging 10-15 miles per hour with frequent gusts above 20 miles per hour, persist across much of the plains area where there are few obstructions to slow them down. To further compound matters, Kansas is also located at a climatic boundary, with frequent intrusions of warm moist air coming up from the Gulf of Mexico, and/or cold dry air coming down from the Canadian plains. These conditions can lead to strong and severe thunderstorms and tornadoes, and occasionally to severe blizzards in the winter as well. Those establishing homes and farms on these Western Kansas plains had to learn to endure this harsh climate in addition to any other hardships they encountered.

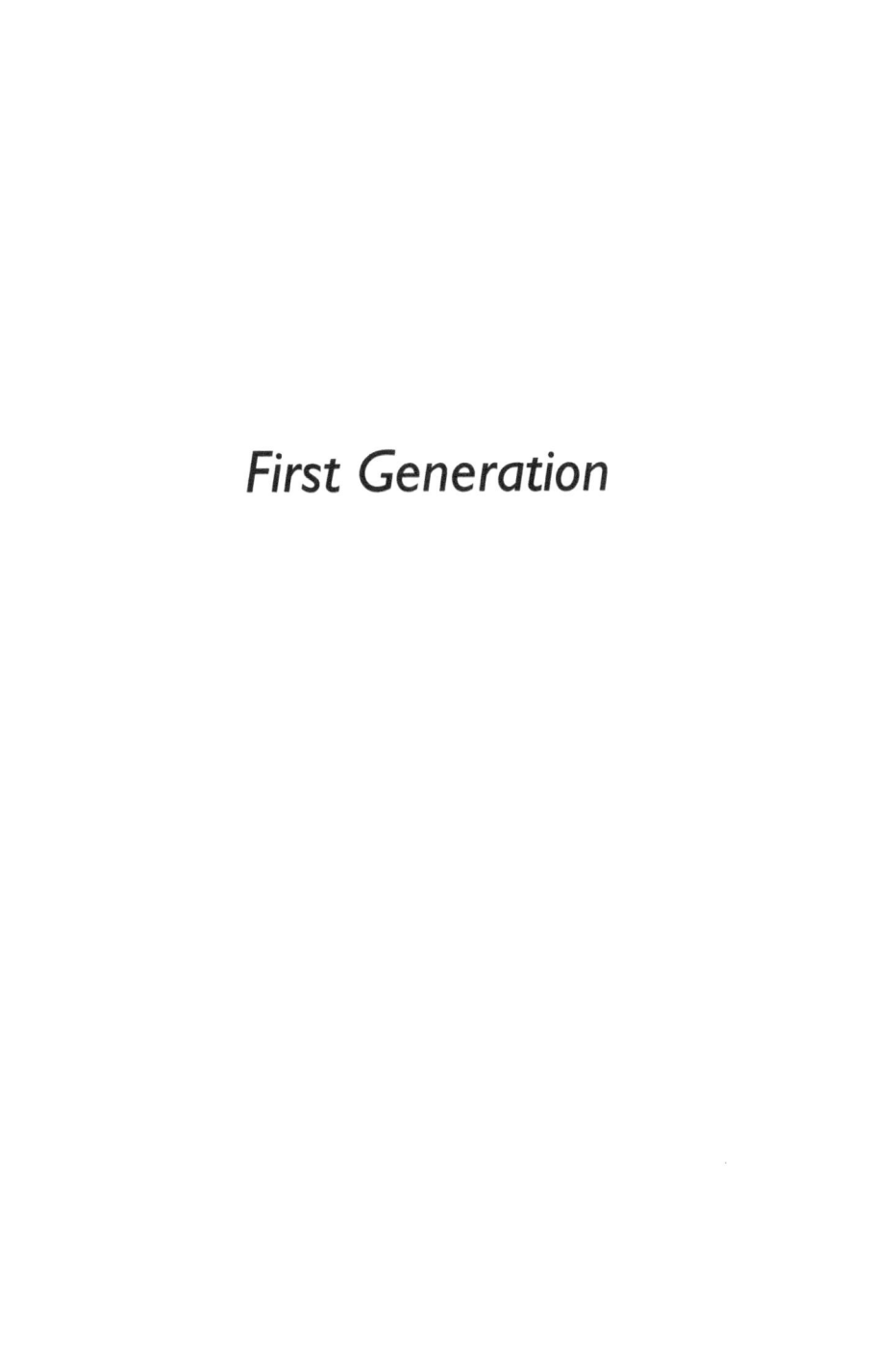

First Generation

FROM PRAIRIE TO VIABLE FARM

The New Frontier—Russell County

Harmon Crawford was excited, and a bit apprehensive. Today, as usual he awoke before dawn and set out to do the morning chores. However, this was not to be a usual day—a radical idea had been gradually gaining traction in his mind, and today he would share this idea with his wife Candus.

Harmon and Candus had been farming on a portion of his father's land in what was then Fishing Creek Township in Columbia County Pennsylvania. However, Joseph Crawford II, Harmon's father, did not have enough land to provide viable farms for each of his sons, and Harmon and Candus were barely eking out a living on the small amount of Joseph's land they were farming. Columbia County was becoming crowded, and land there was too expensive for Harmon to consider the purchase of additional farmland nearby, so Harmon could see that he would have to make other plans for their growing family. His radical idea was to leave this Pennsylvania farm area, the only home they had known since they were born, and head to the frontier where land was cheap or even free.

He and Candus discussed this idea for many days, and as time went by, they became increasingly enthusiastic about making a new start. Neither Harmon nor Candus had ever traveled beyond the borders of Columbia County, but they each harbored an urge to see other parts of the country. This was their opportunity to do so, and after much discussion they decided to set out for the frontier, where by homesteading they could obtain enough land to farm and to put down new roots for their family.

By 1877 they were prepared to start west with their children and a few of their possessions. They hoped to avail themselves of the opportunities available on the new frontier, which by now had progressed further westward to the central part of Kansas. They were largely unaware that that frontier area bore little resemblance to the Pennsylvania land they had previously called home. Annual precipitation levels were far less than in Columbia County, and the near absence of trees meant the visible horizon extended for miles, amplifying the sense of unlimited empty space found on the prairie. Another noteworthy feature of Western Kansas was the never-ending wind, occasionally accompanied by violent storms. The Crawfords would be in for a rude awakening when they reached this alien environment at the frontier. However, both Harmon and Candus were strong-spirited, and were stubborn enough to stick with their decision no matter what difficulties they might face.

Shortly after the birth of their third child, Anna, on July 16, 1877, Harmon, Candus, and their family left Columbia County Pennsylvania with the goal of homesteading in Kansas. They took the train to Kansas City, Missouri, stopping there for nearly two years to earn the money they would need to establish and equip their farm once they had staked their claim.

Harmon's aunt and uncle, Elizabeth (Crawford) Bright and Richard Bright, had moved to Wyandotte County Kansas in 1875 after having previously lived in Columbia County Pennsylvania where Richard worked as a carpenter. Wyandotte County was just across the Missouri River from Kansas City, Missouri, and Richard and Elizabeth helped the Crawfords find a home nearby when they arrived in the Kansas City area. Harmon went to work in the meatpacking plant, setting aside some of his earnings to purchase tools, seed, and livestock for the new farm they planned.

Harmon's younger brother Joseph Franklin Crawford, Candus's younger brother George W. Dildine, and the Samuel Johnson family also left Columbia County at the same time as Harmon and Candus and their children, and all these would-be settlers from Columbia County traveled together to Kansas City. George was a skilled carpenter, trained by his father. He expected to find carpentry work near the frontier. The rest of this group intended to establish homesteads and to farm.

After arriving in Kansas City, George took the train farther west to scout the area where the group hoped to establish their new lives in Kansas. He had heard of a colony of sixty people from two southeastern Pennsylvania counties who had settled along the

Solomon River valley in 1871, in what later became Osborne County. There they had established successful farms and had founded the town of Osborne City (later Osborne), now the county seat of Osborne County. George set out to investigate that general area.

The closest railroad station to that area was the one on the Kansas Pacific line at Russell, Kansas, and George bought a ticket to that station. Train fares at that time were 2-3 cents per mile for the cheapest tickets, a third-class seat on a bench, so a ticket from Kansas City to Russell cost George about $7, about four day's wages for working in a Kansas City meatpacking plant.

Russell was the county seat of Russell County, located directly south of Osborne County. The Kansas Pacific Railroad proceeded nearly straight west from Kansas City before it reached Russell County, a little more than half way across the state. The railroad company built stations in Bunker Hill and Russell within Russell County, and towns developed around both sites.

After considerable political infighting, the people of Russell managed to gain the county seat in 1874, and Russell remained the county seat thereafter. By the late 1870s, Russell was a well-established town with a population of about 800. The bustling settlement sported several grocery stores, drug stores, general merchandise stores, hardware stores, millinery shops, lumberyards, grain elevators, a flourmill, a broom factory, two carriage and wagon shops, several newspapers, and two hotels. Russell also had churches, a fine schoolhouse, and several doctors and lawyers.

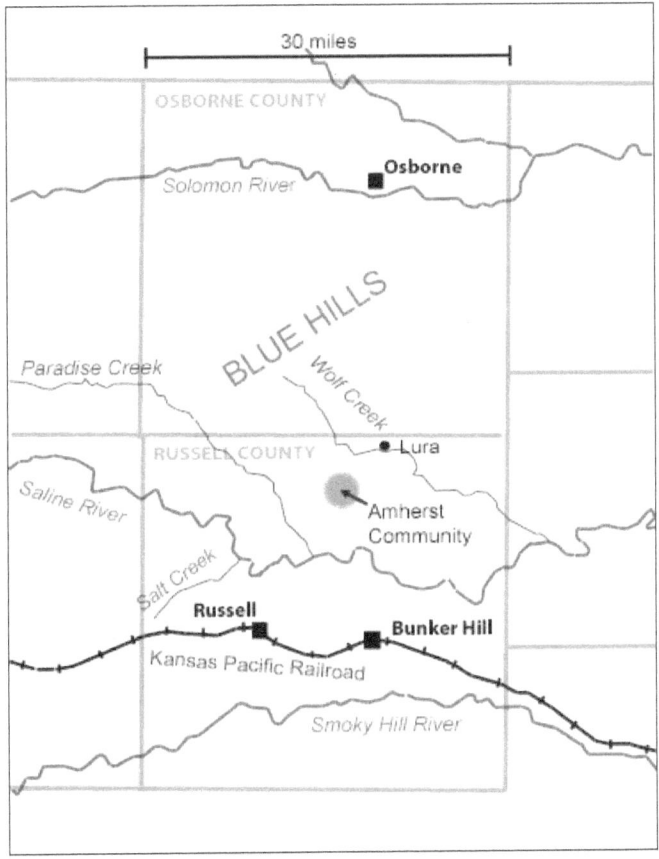

3. Osborne and Russell Counties, (ca. 1880). The town of Luray later arose near the site of the early post office, Lura. The Amherst community existed in 1880, but it did not have that name until a few years later.

By early spring in 1879, after nearly two years in Wyandotte County, the Crawfords were ready to leave Kansas City and move west with their family to establish their homestead. Sadly, Amos Elmer and Sarah Catherine, the two older Crawford children, had died of diphtheria and were buried in the Quindaro Cemetery while the family was living in the Kansas City area, so the family was smaller by this time.

George Dildine had previously familiarized himself with the Russell and Osborne areas, identifying potential homestead spots in northern Russell County. Settlers had already claimed most of the land east of Russell County as well as parts of Russell County itself, but the northern part of Russell County still had homestead land and railroad land available. The spots George found were in the area labelled Amherst in figure 3. The lack of potable river water or deep ground water had so far prevented extensive settlement in this area.

Two rivers cross Russell County from west to east, the Smoky Hill River to the south and the Saline River to the north. The settlers would have to cross the Saline River to get from Russell to the potential homestead sites George had identified. As the name implies, salt springs along Salt Creek made the Saline River downstream from that creek too salty to use for drinking or for irrigation. Much of the deep water in the northern part of the county was also saline. There, one could find fresh water only in the streams or in shallow wells dug fifteen to thirty feet deep to reach the water table.

Nevertheless, the land in the Amherst area would make good farmland, and this was where Harmon, Candus, and the rest of their group planned to select their homestead sites. This land was part of the mixed-grass prairie, an area dominated by the short grasses such as blue grama, sideoats grama, and buffalo grass, but also with a few areas of tall grasses such as big bluestem, little bluestem, and switchgrass. Other plants native to the area included sunflowers, western ragweed, and pigweed. The hot dry climate

24

also promoted the growth of some succulents such as prickly pear cactus and soapweed yucca.

The Kansas Pacific Railroad owned alternate sections of land in the Amherst area, but much of the other land there was still available for homesteaders. The nearest towns were Russell, about 13 miles to the southwest; Bunker Hill, about 11 miles to the south; and Osborne, about 26 miles to the north. The Saline River passed about five miles south of the Amherst area, and Wolf Creek, which flowed with good water at least during times of rainfall, passed about three miles north of that area. Osborne County began a couple miles north of Wolf Creek, and the Solomon River crossed through the northern part of that county. The area known as the Blue Hills spanned most of the land between Wolf Creek and the Solomon River.

The Harmon Crawfords, the Samuel Johnsons, and Joseph Franklin Crawford took the train to Russell, where they met George. Since Russell was a well-established town by the time those settlers arrived, they could purchase needed supplies there.

For Harmon and Candus, those supplies included a pair of horses with appropriate harnesses, a spring wagon with a canvas cover, a plow to turn over the sod, and a harrow to break up the plowed sod. They also acquired seeds for the crops they would be planting, and a supply of food to tide them over until they could produce their own food.

A workhorse cost around $60, a good spring wagon about $30, a one-bottom moldboard walking plow nearly $15, and a harrow about $13. Cooking utensils and supplies were also essential. They also acquired a rifle, shotgun, and ammunition for shooting game or predatory animals. Other things they needed included a sickle or scythe, shovels, a pickaxe, and other hand tools. They planned to buy a milk cow and some chickens and hogs from one of the neighbors, after settling on their homestead.

After a few days of shopping, they had acquired the supplies they needed and had loaded them on their new wagons. The next morning the group of travelers consisting of Harmon, Candus, and Anna Crawford; Joseph Franklin Crawford; Samuel and Lizzie Johnson and their family; and George Dildine set out northeast from Russell toward the potential homestead sites George had identified.

The sky was clear, the sun was bright, and the humidity was so low they could see for miles. George, who had previously explored the route they would follow, led the way. As the group left Russell, the panorama of the prairie engulfed them. A vast sea of grass appeared as flat as the surface of a lake and extended as far as the eye could see in all directions. By squinting, they could spot a few widely scattered dwellings that appeared as tiny islands in this sea. Little else, not even a tree, intruded on this vista.

As they proceeded along the route, they discovered that this "sea" was far from flat and that they were traveling through an area of rolling hills—in fact the teams of horses struggled to pull the heavily laden wagons up some of those hills.

The scene was even more startling when they reached the Saline River, where they found seemingly impassable high and broken bluffs along portions of the river. However, George knew the way and led them some distance along the river until they reached a stretch where they didn't have to descend or ascend any of the bluffs, where the river was shallow enough to ford, and where the grade on the far bank was gentle enough for the teams to climb once they were beyond the river.

After crossing the Saline, they still had a few miles to go before they reached the area George had found for them to consider homesteading. They had to cross Paradise Creek before reaching that area, but the bluffs at Paradise Creek were less imposing than those at the river. With only a trickle of water running in the creek, this crossing was no problem, and by early afternoon, the group had reached the area where they expected to establish their homesteads.

After inspecting the potential homestead sites, Harmon and Candus selected as their homestead the northeast quarter of section 34, township 11 south, range 13 west of the Sixth Principal Meridian, in the area eventually known as the Amherst Community. The shorthand representation for this would be NE¼ 34-11S-13W, or NE¼ 34-11-13, or possibly even just northeast quarter of section 34.

Joseph Franklin Crawford and the Samuel Johnson family (Samuel, his wife Lizzie, and sons Levi and Charles) also selected homestead sites in the Amherst community, a little east of the site Harmon and Candus had selected.

George Dildine had already set his sights about 26 miles north of the Crawfords in Osborne City, Kansas, where he had previously met Emma Jemison. George and Emma would marry in September 1879, and would live in Osborne City for a few years, while George worked there as a carpenter. Osborne City had become the county seat of Osborne County, and was growing rapidly, providing plenty of work for carpenters at that time.

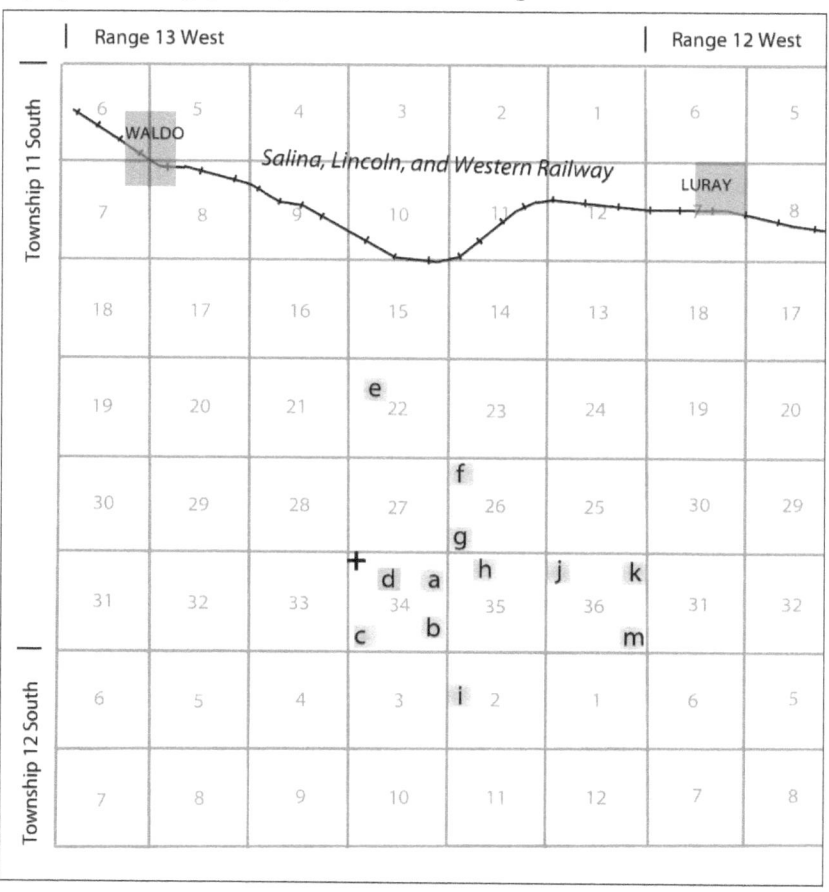

4. Some Amherst area families and dates they established their homes. a: Harmon Crawford (1879); b: Humphrey Leighton (ca. 1878); c: Dan O'Connor (1878); d: Thomas Beatty (1880); e: Rudolf Whitman (ca. 1873); f: C. S. Wycoff (ca. 1878); g: Joseph King (1878); h: Roy Crawford (1913); i: Jacob Bean (1881); j: Benjamin Bratton (1878), k: Samuel Johnson (1879); m: Alex Hampl (1908); +: Amherst church (1906) and school (ca. 1893). The Salina, Lincoln, and Western Railway and the towns of Luray and Waldo all appeared in 1887.

There were already some settlers in that Amherst area (figure 4). The first of these were Mr. and Mrs. Rudolf Whitman, who had homesteaded there in the early 1870s while the Native Americans and the buffalo were still plentiful in these parts. The Whitman family had also come from Pennsylvania, and this may have influenced Harmon and Candus's decision to select this area for their homestead.

Others already settled on section 34 when the Crawfords arrived were the Humphrey Leighton family in the southeast quarter and the Daniel O'Connor family in the southwest quarter. Humphrey Leighton and his wife Sarah had come from Iowa with their six children in about 1878. Daniel O'Connor and his wife Catharine had arrived from Iowa, also in 1878, with their three children.

Joseph Harrison King, a Civil War veteran, and his family were living on the southwest quarter of section 26, diagonally northeast from and touching the northeast corner of the Crawford land, having arrived from Indiana in 1878. Joseph and his wife Mary had four children. Alfred King, Joseph's nephew, had also arrived in 1878 and had homesteaded the southeast quarter of that same section. However, Alfred sold his land shortly after he received his patent in 1883 or 1884, and after that, no one resided on that quarter.

No section-line roads or other roads existed in the Amherst Community when the Crawfords arrived. Folks just went wherever it was easiest to travel. One main route from Russell to Osborne passed through the King farm (letter g in figure 4). Every other day, a stagecoach brought the mail to a Post Office, known as the Wyckoff Post Office, in the northwest quarter of section 26.

Planting Roots

Once Harmon and Candus had selected their homestead site, they parked their wagon, and along with Anna, they began to explore the area. A ridge ran north to south about 300 yards west of the eastern edge of what would become their property. The Crawfords decided they would locate their home on the eastern slope of that ridge so the ridge would provide a little shelter from the almost constant westerly wind. After climbing to the top of the ridge, they had an unrestricted view of all of section 34 and many of the surrounding sections. Most of that vista was rolling buffalo grass prairie, but they could also see home sites of several other settlers who had arrived earlier and had already built houses or other structures. That night the tired family retired to their wagon to sleep, with visions of what this property could eventually become once they set to work on it.

When Harmon, Candus, and Anna first arrived on their land in the spring of 1879, their highest priority was to provide a reliable supply of water for themselves and their livestock. Harmon immediately set to work digging a shallow well. Fortunately, at that time of year the Crawfords could get some water from a small spring on railroad land a little more than a half mile north of their homestead. Until Harmon completed the well, a family member had to make this trip daily unless they were fortunate enough to collect some rainwater instead.

Harmon hand-dug their well to a depth of about 16-20 feet, which was enough to collect a supply of potable ground water that gradually seeped in from the surrounding water table. Initially, the family used a bucket on a rope to haul up water from this well. Later Harmon installed a hand-pump and eventually a windmill-operated pump. This well served them for many years, and except during times of extreme drought, it provided an adequate supply of water for the family, a limited number of livestock, and a small garden.

After completing the well, the next priority was shelter for the family. Initially they lived in their spring wagon until they could create a more permanent abode. The spring wagon had a canvas cover that gave them some protection from the elements, but as soon as he finished the well Harmon immediately began to build a dugout for the family to live in, locating it in the east side of the ridge near the eastern edge of their property, about a hundred yards north of the well.

Harmon would use sod for some of the walls of the dugout, so the first step was to make some sod building blocks. Harmon used his plow, a moldboard plow like the one in figure 5, to cut some of the sod from the surface of the hill.

The primary components of this plow were the plowshare and the moldboard. The front edge of the steel plowshare made a vertical cut in the soil to define one edge of the furrow, and the bottom edge of the plowshare made a horizontal cut from the previous furrow to this vertical cut, releasing a strip of sod. The plowshare lifted this strip and the steel moldboard rolled it over to fall upside down beside the furrow. The team of horses pulled the plow, while Harmon walked behind and guided the plow and the horses.

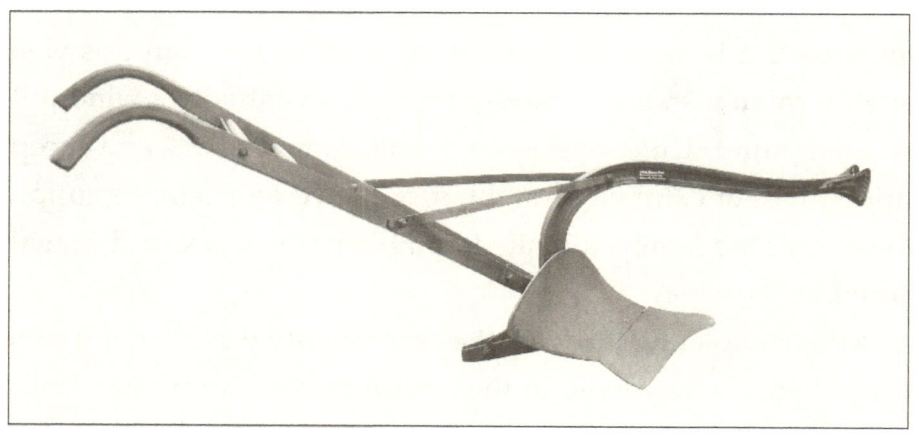

5. Walk behind horse-drawn moldboard plow, an implement introduced by John Deere in 1837.

Following this operation, Harmon cut the sod into lengths suitable for building blocks and set each block aside. Using a shovel and a pickaxe, he dug into the side of the hill to create the back wall, floor, and parts of the side walls of the dugout. He completed the side walls and built the front wall by stacking the blocks of sod he previously had set aside, taking care to leave space for a door and a window.

Next, he collected wood and brush to provide a structure to support the roof. There were no trees on the Crawford homestead land to provide wood for the dugout roof support, but Harmon found enough small trees and brush along the often-dry gully that ran through the nearby railroad land.

Once the support structure was in place, Harmon completed the roof with grass and sod and plastered any remaining holes in the roof and the sod walls with mud to seal them. There are no photos or even any detailed descriptions of the Crawford dugout, but the completed dwelling probably looked somewhat like the one shown in figure 6.

6. Kansas sod dugout. (ca. 1870-1890)

Like other settlers who came from small farms in the eastern states, the Crawfords were used to a largely self-sufficient farm economy that depended primarily on raising hogs, and on growing corn as feed for the hogs, oats or barley as feed for the horses or mules, and perhaps a little wheat for flour or as a cash crop.

The hogs ate the bulk of the corn, and the family ground the remainder of the corn to make cornmeal. The family would butcher some of the fattened hogs, and sell the rest. Cured pork made up a significant part of the farm family diet, and corn bread spread with sorghum molasses was a standard part of most meals.

They ground the wheat to make flour, adding a little variety to the farm diet, or else bagged it and took it to a market for sale. The farm family usually owned one or more milk cows and a flock of chickens. These provided milk, eggs, and occasional chicken dinners for the family.

The Crawfords initially attempted to apply a similar economic model on their homestead. It was spring when they arrived on their homestead, and they planted spring crops such as corn, barley, and spring wheat as quickly as possible after their arrival. They also obtained a milk cow from one of the neighbors. The cow and the horses could feed themselves by grazing, at least until winter approached. Once they had harvested their grain, they could use some of that grain to feed some chickens and hogs, which they then acquired from their neighbors.

With a team pulling his one-bottom walking plow, Harmon could break out about two acres of sod per day, making it a painfully slow operation. Harmon and the team then had to pull the harrow over all the plowed land to break up the plowed sod into finer pieces suitable as a bed for the crop. This went faster than the plowing, but it still took time. Harmon only had time to prepare a few acres before he had to plant the spring crops that first year.

Once he had plowed and cultivated the ground, Harmon planted wheat and barley by manual broadcasting, strewing the grain by hand evenly over the soil and covering it up by tamping it in by foot or other means. This was a tedious and inefficient process, which left many of the seeds planted too shallowly to grow properly. Also, some areas ended up with too many seeds growing so that none of them grew well.

At harvest time, late in the summer, Harmon used a hand sickle or a scythe to cut the grain stalks. Then, with help from other family members, he gathered the reaped grain stalks into bundles or sheaves and loosely tied them with a piece of twine or a bit of straw, standing several bundles together in a shock to hold each other up and keep the ears of grain off the ground to dry out.

After the shocks were sufficiently dry, Harmon used a stripper, an implement shaped like a large comb with the teeth pointing upward, to remove most of the straw from the heads of grain. He then spread the heads of grain out on the threshing floor, a hard surface of carefully leveled and packed earth, and the family took turns beating the heads with the flail to separate the grain from the chaff and straw. After beating out the grain with the flail, they carefully raked the straw away and winnowed the grain and chaff by tossing the mixture up in the air to let the wind blow the chaff away while the grain fell back to the threshing floor.

They could thresh an average of about eight bushels of wheat or barley a day with the flail. A good wheat crop in those days would be about 15 bushels per acre, and each acre of wheat or barley required 50-60 hours of Harmon's labor to plow and harrow the soil, hand broadcast the seed, reap with a scythe, and thresh with a flail. Given the other priorities for his time, in that first year Harmon did not have the time to plant and harvest more than a few acres each of these grains.

That first year, Harmon also plowed a couple acres for planting corn. He used a hoe to pulverize and shape the plowed and harrowed sod into small hills, placing three or four seeds in each hill. The corn ripened in the fall, and then Candus and Anna helped with handpicking the ears of corn. The family needed to put in 30-40 labor-hours per acre doing the plowing, harrowing, and hand planting and harvesting to produce ears of corn.

The Crawfords used the barley as supplementary food for the horses and the milk cow they had acquired, and maybe to brew some beer as well. Later, they took the corn and wheat to a mill where the miller ground it into meal and flour for the family, saving a little to feed the chickens and selling any remaining corn and wheat to provide needed cash.

Harmon also plowed a small area near the well for a garden, and there the family used hoes to break up the sod and the resulting smaller clods so the soil would be suitable for planting. Candus and Anna planted and tended this garden, likely growing an assortment of pumpkins, squashes, beets, carrots, watermelons, herbs, and perhaps even asparagus. Some of these needed frequent watering to survive the harsh Kansas summer. Their garden also included about a quarter acre of potatoes. This garden provided them with an important source of food, some of which they preserved for the coming winter.

The dugout, although providing better shelter and more living space than the spring wagon, was not a pleasant place to live, and the roof or walls were liable to collapse during a heavy rain. By late summer, after they had completed the labor-intensive process of harvesting their wheat and barley, the oppressive summer heat, the constraints of living in the dugout, and the ever-present wind on the Kansas prairie left Candus depressed and homesick. Harmon initially didn't know what to do about this, but then he had an inspiration. He drove with Candus in the wagon to visit the Rudolf Whitman family, who had been in the area longer and had a real house with a wood floor. When Candus first set foot on that wooden floor, she was finally convinced that it really would be possible to have something better than a dugout with a dirt floor to live in here on the prairie. That thought helped her to recover from her homesickness and depression, and sustained her until the family achieved that goal. This episode convinced Harmon that it was imperative to build a proper house as soon as possible.

Because they had no trees and could not afford to buy much lumber, building a house of wood was out of the question. However, others in the neighborhood had already started building houses of limestone blocks, and this was what Harmon set out to do. Limestone layers outcropping in the area were six to eight inches thick, just about right for building blocks, and Harmon began using any spare time to quarry limestone blocks for the house. The limestone was relatively soft until exposed to air, and

Harmon could quarry it easily using hand drills to drill a few holes and then hammering on "feathers and wedges" placed in those holes to split the stone along the desired line. Finally, he used a hammer and a stone chisel to square up the ends of the stone blocks so they would fit tightly together. George Dildine was a builder by trade, so it is likely that he helped Harmon with this task, and Joseph Franklin Crawford may have helped as well.

Harmon built the Crawford stone house a little to the south of the dugout. The house had three rooms to start with, with a cellar extending under two of them. A few large pieces of limestone covered with dirt provided the roof for the cellar as well as the floor for the rooms above it. The cellar served as a cool storage place for food and as a safe retreat for the family whenever one of the frequent violent storms occurred. A mortar made from slaked lime, a lime powder made by burning broken pieces of limestone in crude kilns, filled the spaces between the blocks and held the stone walls of the house together. Wesley Van Scoyoc, an early settler, had built a lime kiln on Wolf Creek, a few miles northeast of the Crawford homestead, and that was where Harmon got his lime.

Initially, only the walls of this house were stone; the floor was dirt and a log or two held up the roof made of brush and sod. The roof required long strong logs for support, so Harmon and Joseph Franklin drove the wagon to Wolf Creek, the closest place to find any trees that large. There they selected and felled the trees to provide the logs they needed, and hauled those logs back to the homestead in the wagon. The house also had a fireplace, which provided for warmth and cooking. Later, Harmon built on a long kitchen and another bedroom, and he eventually added a wooden floor and a proper roof with wood framing and wood shingles.

Harmon had to take some time away from the building to break out more land and to plant some winter wheat. He also built some structures to provide shelters for the livestock during the coming winter. He constructed these of poles over-laid with brush and roofed them with prairie hay or sod.

Despite all these tasks, Harmon somehow also found time to use a scythe to cut prairie grass for hay. He left this grass on the ground to dry for a few days before raking it into piles, and then he loaded it on the wagon with a pitchfork. This was backbreaking work, and he vowed to purchase a horse-drawn mower as soon as he could afford it. He prepared several wagonloads this way, and then stacked the hay for use as feed for the livestock in the winter. This provided the livestock with better quality nutrition than simply allowing them to dig through snow in the winter to find dried grass.

While Harmon was doing the digging, plowing, planting, and building, Candus and Anna dealt with many other tasks including cooking, sewing, laundry, milking the cows, churning the cream into butter, gathering the eggs, tending the garden and the potatoes, and helping with the harvesting of the crops.

They also had to find the fuel to use for cooking and for heating. They scoured the surrounding prairie for buffalo chips, trying to keep a good supply on hand to burn for fuel. The buffalo, which had once been so plentiful across the prairie, left behind an abundance of dried buffalo manure in the form of patties commonly known as "buffalo chips." These were a welcome source of fuel for cooking and heating for the early settlers, especially in the prairie areas where there was little or no wood to burn.

Until the family was able to harvest some of the grains and vegetables they had planted, they had to survive mostly on the food they had brought with them when they arrived. Whenever possible they supplemented this with whatever wild game or birds they were able to shoot, including prairie chicken, quail, and rabbit.

The rainfall in Kansas was above average in 1879, and the spring crops Harmon had crudely planted provided an adequate harvest. After harvesting those crops, Harmon took an overnight trip to Russell, driving the wagon loaded with some of the grain and some of the hay he had cut. There he had some of the grain ground into flour and sold the rest of the grain and the hay. Harmon used the proceeds from those sales to purchase necessary items for the family, including beans, sowbelly, material to make clothes for the family, additional tools and supplies, and doors and windows for the new stone house.

Even with all these interruptions, the family was still able to move into the completed stone house in late December of 1879. This was fortunate, because Harmon and Candus's fourth child, Minnie Almerta (usually referred to as Mirt), was born January 27, 1880, in the stone house on the homestead. Roy Alfred Crawford, Bertha Alice Crawford, and Charles Clark (Charlie) Crawford were also born in that stone house on January 29, 1882, December 27, 1883, and October 22, 1885 respectively.

7. Rendition of the original Crawford stone house after the addition of the new roof. Viewed from the southeast.

A Difficult Decade

Shortly after the birth of Mirt, the Thomas Beatty family showed up at the Crawford homestead. The Beattys arrived from Crawford County Pennsylvania in February 1880, and intended to homestead on the quarter section just west of the Crawford farm. Thomas Beatty was 49, his wife Jane was 48, and they already had ten children, many of whom were old enough to help with the farming or to homestead on their own.

Even though their new house was not large, the Crawfords invited Thomas, Jane, and the youngest of the Beatty children to

41

stay with them until early March. That was when the Beattys, with Harmon's help, finished building their sod house on their new homestead land.

One day while the Beattys were still staying at the Crawfords, both families were sitting down to dinner in the new stone house. Thomas Beatty, a deeply religious person, was praying before everyone began to eat. His prayer went on at some length, and while he was praying a big patch of the brush and sod ceiling plopped down onto the table, the food, and the diners. A portion came to rest on Mr. Beatty's bald head, but he kept right on praying, even though Harmon's loud cussing of that ceiling soon drowned out Mr. Beatty's prayer. Laura (Crawford) Luder, who recounted this tale, told to her by her parents, added, "I guess my father had a right to swear as he was furnishing the food."

Eventually they were able to clean up the mess and to salvage enough of the food to continue with their dinner. Later, the Crawfords were able to enjoy many a good laugh about this episode. The indelible image of Harmon cussing a blue streak at the ceiling as Mr. Beatty solemnly continued to pray while wearing a portion of their sod ceiling on his head, remained an important part of the family lore.

The arrival of the Beattys completed the settlement of section 34, as they claimed the last of the available land on that section. The 1880 US Census lists eight Humphreys, six O'Connors, four Crawfords, and ten Beattys living on established homesteads on all four quarters of section 34. Less than ten years earlier this land had all been virgin prairie supporting only buffalo and other wildlife and a few nomadic Native Americans. Now the buffalo and the Native Americans were gone and this section of land supported four families comprised of 28 people plus their livestock.

Life on the Kansas prairie was hard, and those trying to establish roots there often worked at other jobs when they could to earn a little more cash. Harmon's brother, Joseph Franklin Crawford, had been working for the railroad during the winter to supplement the meager earnings from his homestead. He died in a train wreck near Santa Fe, New Mexico on December 25, 1880. This was a tragic loss for the Crawford family.

The early settlers were never far from danger, even on their own farms. Rattlesnakes were one of the menaces found on the prairie, and in the early 1880s, one bit Anna Crawford on her foot. Anna was about five years old at the time. Harmon tied a tourniquet around Anna's leg, and according to Laura, "Then they poulticed it with melted soap and soda and gave her lots of whiskey. She got all right."

This anecdote illustrates how resourceful the settlers had to be to handle medical emergencies. It also draws attention to the fact that whiskey was an important medicinal item in the early days. The Crawfords brought a case of McHenry whiskey, produced in Columbia County, with them when they took the train west from Pennsylvania. They intended to use this whiskey for medicinal purposes, but apparently it sometimes found other uses. Family lore holds that one day when Mr. Rudolf Whitman came staggering home, his wife's first exclamation was, "My God, he's been to the Crawfords!"

Laura related another example that epitomizes the medical resourcefulness and the kind of physical and mental toughness required to survive those early days on the prairie. During their early years on the homestead, Candus developed an abscess in her throat. Since there were no doctors readily accessible, she took her own initiative to deal with this issue. When she could not stand it any longer, she had Harmon sterilize a straight razor. Then, using a mirror to see what she was doing, she painstakingly inserted the razor into her throat and cut the abscess open to drain. There is no record of whether she followed up this surgery with any applications of their medicinal whiskey.

Many travelers would stop at the King, Crawford, and Beatty homes while passing through. Occasionally they arrived at inopportune times. The grass, including that in the sod roofs of many houses, was full of fleas and other bugs, necessitating occasional breaks to remove the fleas from one's body. One day in 1882 Candus was half undressed picking fleas off her clothes when a man she did not recognize walked up to her door. Startled and embarrassed, she dressed, answered the door, and came face to face with Jacob E. Bean who introduced himself. Jacob said he was planning to establish his homestead in the area. The Crawfords took him in, and he stayed with them while Harmon helped him build a sod house on his homestead quarter about two miles south of the Crawford homestead.

The high winds on the Great Plains led to frequent dust storms, even before the settlers came and disturbed the prairie with their plows. As more settlers arrived and began plowing up the prairie grass and planting wheat or corn in its place, this left bare soil exposed, making it even easier for the wind to lift the fine particles of dirt into the air.

April of 1880 brought the first dust storm the Crawfords had encountered after arriving on the prairie, and this was a particularly bad one. One account of that storm noted, "The wind was fierce and the atmosphere was blindingly full of sand and dust, giving it the appearance of a yellowish impenetrable fog." This storm persisted off and on throughout most of the month of April. Less severe dust storms continued to occur throughout the early 1880s, and the dust storms in April and May of 1883 were nearly as severe as the one in 1880.

The mid-1880s were a period marked by exceptionally nasty weather. The eruption of the Krakatoa volcano in Indonesia late in August 1883 had a pronounced effect on worldwide weather, darkening the skies with a cloud of sulfurous ash that circled the globe and produced chaotic weather patterns that continued into 1888. The first major weather effect across the Great Plains attributed to the Krakatoa explosion was a far-reaching severe cold spell coupled with blizzards that occurred during the winter of 1884-1885. In late December 1884 temperatures dropped to 20 to 30 degrees Fahrenheit below zero, and they remained at those levels well into February 1885. Howling winds and drifting snow accompanied these frigid temperatures.

The Crawfords vividly remembered one horrible consequence of this blizzard. Their neighbor Catharine O'Connor gave birth to her sixth child during this blizzard in January 1885, but shortly thereafter Catharine died due to complications from the birth. The ground froze so solidly that her husband Dan was unable to dig even a shallow grave to bury her. The blizzard prevented the family from leaving the farm, and they had no other option than temporarily to bury her body in a snowdrift. When the thaw finally came in March, Dan loaded her body in their old wagon and with the children drove the wagon back to Iowa, where they buried Catharine. Dan and the two older boys, Francis and Henry, then drove their wagon back to Kansas to resume farming on their homestead, leaving his three-month-old baby Lizzie, his youngest son Leo, and his daughters Mary Catherine and Ellen to be raised by family in Iowa.

Although the winter of 1884-1885 was bad, worse was yet to come. The Plains Blizzard of late 1885 dumped heavy snows on Kansas, with the accompanying high winds leading to drifts piled ten feet high. The Kansas Blizzard of 1886, occurring in the first week of January 1886, reportedly froze 80 percent of the cattle in the state. Nearly 100 Kansans froze to death during this storm as well. The storm paralyzed businesses and rail traffic for weeks. A train, stopped by the snow in western Kansas, froze to the rails and each car had to be separately uncoupled and broken loose from the tracks after clearing the snow.

The Great Plains Blizzard of late 1886 brought snow that started in mid-November of 1886 and did not stop for a month. The Great Plains Blizzard of 1887, a 72-hour blizzard that began in January 1887, followed almost immediately, depositing an additional 16 inches of snow whipped by intense winds. Temperatures dropped to 50 degrees below zero Fahrenheit. Most cattle that did not die from the cold soon died from starvation. When spring arrived millions of cattle were dead, most rotting where they fell. Many ranchers went bankrupt and others just quit and moved back east.

Fortunately, the Crawfords had planned for adequate shelter, food for their livestock, and fuel for their fireplace, so they managed to make it through these blizzards without suffering major damage to themselves or their farm animals. The blizzards occurred in the winters, and at that time of the year the wheat and any other crops planted in the fall were dormant, and the Crawfords had not yet planted corn or any other spring crops. Hence the blizzards caused little or no direct damage to their crops. However, every day of each blizzard Harmon had to bundle up and struggle through the freezing winds and the drifted and blowing snow to feed the livestock and to make sure their shelter was adequate. He wore his long woolen underwear and donned as many layers of trousers and shirts as he could manage, before finally putting on his heavy coat.

Despite these storms, the early 1880s had been years of plentiful moisture, and the crop and hay harvests had been good. Harmon was able to save enough cash to purchase a wagonload of lumber in Russell, and he used that lumber to build a barn. He got George Dildine to help with the design and construction. This barn included stalls and other spaces for the livestock, bins for storing grain, a small area for a threshing floor, and a space for storing hay. The Crawfords needed all the sheltered space provided by the barn, especially during the blizzards in the mid and late 1880s.

The family found that conditions on the Kansas plains were very different from the conditions in Pennsylvania—a hotter, drier climate confronted them, there were no streams or rivers to supply water for the livestock, and they faced more violent and destructive storms and an incessant wind. They gradually adapted their economic model to place less emphasis on corn and hogs and more emphasis on wheat, a more reliable crop under the conditions found in western Kansas.

By 1885, they had five horses, four milk cows, and four hogs. At that time, they were growing 19 acres of winter wheat, five acres of spring wheat, 10 acres of barley, half an acre of potatoes, and had 300 bushels of wheat stored in the grain bins. They had cut 15 tons of hay in 1884, made 200 pounds of butter, and sold $30 worth of poultry and eggs that year. Wheat was worth about $0.50 per bushel at that time, so the 300 bushels of wheat they had stored represented a nest egg of about $150. An average wheat yield at that time was about 15 bushels per acre, so the wheat they had planted would represent another roughly $180 of expected income for the year.

During the last few years of the 1880s, Harmon began fencing some of the Crawford land. At the time the Crawfords arrived, there were almost no fences in the Amherst community. Consequently, all the land was essentially free-range land where cattlemen could let their cattle wander or could herd them to graze wherever they wanted. The grazing cattle often destroyed some of the farmers' crops, so farmers began to put up fences around their land to keep the cattle out. This was the purpose of the first fences Harmon erected—they served to keep livestock belonging to others, particularly livestock of nearby ranchers, from intruding on the Crawford land. Later, Harmon also erected some fences to keep his own cattle confined.

Since there was no wood available for wooden posts, other settlers in the area had been using stone posts and barbed wire for fences, and Harmon followed this example. He had to quarry his stone posts from the limestone outcroppings, similar to the way he had quarried the stone blocks for the house. However, a stone post had to be sufficiently long, usually about six feet, so one typical post would weigh nearly 400 pounds.

Harmon used a team of horses to drag each post to its desired position. He would then dig a hole about two feet deep, lever the post into the hole, and carefully fill in and tamp the dirt around the post to hold the post in place. Whenever possible, he got a neighbor to help with this task. Once he established a line of posts, he strung several strands of barbed wire between the posts, making a secure fence to keep the livestock in place. Building a stone post fence was a tedious process, requiring many hours of work, and Harmon took great pride in making each fence line as neat and straight as possible.

In 1886 the Salina, Lincoln, and Western Railway Company began construction of a rail line heading northwestward from Salina and passing through the northern part of Russell County (figure 4). This rail line was eventually 108 miles long and ended just west of Plainville, Kansas. The Union Pacific Railroad system absorbed this rail line in 1898. This branch railroad greatly facilitated the transportation of crops produced in the northern part of the county, and led to the development of the small towns of Luray and Waldo that largely replaced Russell as the trading centers for this area.

Civil War veteran Jonathan Wesley Van Scoyoc founded the town of Luray. In the winter of 1870-1871, he traveled from Ohio to the East Wolf Creek valley in north central Kansas, where he filed a homestead claim along Coon Creek close to where it empties into Wolf Creek. This was in the northern part of what became Russell County, and his was the third homestead claim filed in that county. The creeks near his claim provided enough trees for him to build a log cabin, and that was the first of the few log cabins ever built in Russell County. Having built his cabin, Van Scoyoc went back east to get his wife and young son, and the family set up housekeeping in the log cabin in the spring of 1871. Jonathan raised sheep there, and operated a lime kiln north of the cabin. He became active in the newly organized Russell County government and was elected as coroner in 1872. He retained that position until he won election as a county commissioner in 1883.

Other settlers attracted by the good soil and the plentiful water in the Wolf Creek valley soon joined Van Scoyoc. In 1872, the government established a post office nearby, with Captain John Fritts, another Civil War veteran who had accompanied Van Scoyoc, as the first postmaster. Local legend has it that a heartsick young traveler passing through after a failed love affair suggested naming the post office and nearby school "Lura" for his lost love.

When the railroad branch arrived in 1887, the company chose some of Van Scoyoc's land as the place where it would establish a depot, and a town developed around that spot. The company named the new town Luray after the original Lura. The *Luray Headlight* newspaper, started just before the railroad arrived in 1887, extolled the virtues of land in Luray. Mr. Van Scoyoc took advantage of the ensuing land craze and became a land agent for the railroad, selling land in and near the new town of Luray.

Initially, the railroad extended only about six miles west of Luray to where the company had placed a depot, and had created a railroad wye in the tracks so the locomotives could turn around for the trip back east along the line. The railroad company then transferred about 280 acres of its land there to the Union Town Company, which platted the land to form the town site of Waldo and sold the lots to business and homeowners. Waldo was officially organized as a town in 1888.

Respite in Kansas City

By late 1888, the Crawfords had lived on their homestead for almost ten years and the crops had been good over most of that period. However, they were physically and emotionally exhausted as they had struggled to survive several of the worst blizzards in the history of the country, numerous dust storms, and years of unrelenting wind and extremes of temperature. The past summer had been dry, and by early 1889, it appeared that they would produce only poor crops for that year and that they were entering a period of drought. Furthermore, Candus was pregnant again. These facts prompted them to make the difficult decision to move back to Kansas City for a while. There Candus could have better medical attention, and Harmon could find work to build up a reserve of funds for their intended return to the homestead.

On March 26, 1889, Harmon presented the necessary proof to the Russell County Clerk's office in Russell that he had satisfied the requirements for ownership of the Crawford homestead quarter, and he obtained a Receivers Receipt temporarily conveying ownership of this land to him. He could have done this several years earlier, since he had completed the improvements and period of occupancy required by the Homestead Act by the mid-1880s. However, he didn't have to pay any property taxes on the homestead land until it officially belonged to him rather than to the government, so it had been convenient to wait as long as possible before making the final application for ownership. The impending trip to Kansas City finally forced this action.

8. Harmon Crawford Homestead Patent, with the signature of Benjamin Harrison, President of the United States of America.

Shortly after obtaining their Receivers Receipt, the Crawfords departed for Kansas City. Harmon drove the old spring wagon and Anna and Roy went with him. The wagon had a canvas cover to provide shelter during the trip. A pair of mules, Jules and Jim, pulled the wagon containing most of their important belongings, including their plow. They had sold the horses and the remaining livestock.

Their neighbor Dan O'Connor took Candus, Mirt, Bertha, and Charlie to Russell, where they boarded the train to Kansas City. Neighbors who remained in the area rented and farmed the Crawford land while the Crawfords were away.

9. The Harmon Crawford family in Kansas City. Back row: Anna, Harmon, Roy, Mirt, Charles; Front row: Laura, Bertha, Candus. (ca. 1892)

Harmon and family once again found a place to live in the Wyandotte County Kansas outskirts of Kansas City. Harmon began again working in the Kansas City meatpacking plant, and did some gardening near there as well. The mules, wagon, and plow came in handy for the gardening work.

Candus gave birth to their eighth child, Laura Perl, in Kansas City, Kansas on July 14, 1889. The children that were old enough attended Quindaro grade school, a one-room schoolhouse near the former town of Quindaro in Wyandotte County, Kansas just outside Kansas City, Kansas. On July 3, 1890, Harmon received the official patent deeding him the land they had homesteaded.

54

Return to the Farm

The severe drought on the plains lasted from 1889 to 1896, so it was fortuitous that the Crawfords left when they did. By 1895, there were signs that the drought might be letting up, and there had been no severe blizzards for several years, so Harmon and Candus decided that it was time to return to the homestead. The Crawfords with all six of their children were back on their homestead by March 1, 1895, having spent six years in Kansas City.

The period from 1877 to the late 1880s had been wetter than usual, leading to unrealistic expectations of crop productivity that contributed to the rapid settlement of western Kansas. The settlers had planted corn and wheat and had reaped excellent harvests, and many settlers and land speculators promoted the idea that this was an easy place to make a living. However, by 1889 the cycle had shifted to drought. The droughts of 1889 through the early 1890s were particularly severe, and wheat and especially corn (which was less drought-tolerant) yields fell off. Many who had remained on their farms faced starvation and some simply walked off the land.

Some of the Crawford neighbors who tried to ride the drought out on their farms did not fare well, and the population of section 34 shrank from 28 in 1880 to 13 in 1900. Of the four homesteads in section 34, only two of the original owners were still in place by the end of the 1890s. Dan O'Connor received the patent for his homestead on November 11, 1889. However, the drought in the early 1890s left him in debt, and on April 24, 1899, he sold his 160-

acre homestead to the Agricultural Land Company and left the area. Dan had been a good friend to the Crawfords and they would miss him.

Humphrey Leighton obtained the patent for his homestead on April 29, 1882. The Leighton family then moved away, and after a series of transactions, the person who finally owned the property failed to pay the property taxes. On September 4, 1894, Russell County held a tax auction to sell the land to recover the back taxes. The county received no bids adequate to recover the taxes, with the net result that on August 15, 1896 the county sold the 160 acres of land to Thomas Beatty for the four-years-worth of back taxes, costs, and interest that had accumulated to that time, an amount totaling to $35.18. On August 30, 1898, Thomas Beatty and his wife sold this quarter section to their son Zenas Beatty for $60.

Joseph King, the Crawfords' close neighbor in section 26, died in 1892 while the Crawfords were in Kansas City. His death at the young age of 48 may have been due in part to damage to his health incurred during his participation in the Civil War. However, his wife Mary was still living on their homestead, and five of their children were still at home with her. The oldest remaining child was Harry at age 17, and he began handling most of the King farming with help from Mary and the other children.

In 1897, Harry King married Sadie Beatty, Thomas Beatty's youngest daughter. Mary King and the remainder of her children who were still at home moved into Luray at that time, leaving Harry and Sadie to live in the house on the Joseph King homestead in section 26. They remained there for many years, raising eleven children, all born between 1898 and 1916.

Soon after the Crawfords arrived back on their farm, the Kansas Census showed that they had two horses, two mules, four milk cows, three other cattle, and one hog. They had 60 acres under cultivation, and 60 acres fenced in. They had planted 45 acres of corn and two acres of potatoes, but had not planted any other grain crops by the time of the census. They had no grain or hay on hand. They also had five apple trees and one cherry tree. Thus, when they returned from Kansas City they were not exactly starting over, but they needed to accumulate an emergency supply of grain and hay as quickly as possible, and they would have to use some of their Kansas City earnings to acquire a new set of farm implements.

Perhaps they should have spent another year in Kansas City, since the dust storms came back with ferocity in 1894 and 1895; dust from the Great Plains filling the air to points well east of the Mississippi River. Nevertheless, the Crawfords were back, revitalized by their stay in Kansas City and anxious to begin anew to farm the land they loved and had worked so hard to make their own. The years of 1896 through 1898 were considerably better, but the dust storms returned with intensity throughout the spring months in 1899.

Even when there were no dust storms, the wind would blow day and night. Laura noted that many times the Crawfords would have to brace their south door with a piece of 2x4 construction lumber, driving a nail in the door and a nail in the floor and putting the piece of 2x4 between those nails. The wind usually blew from the southwest and often kept it up for three days or longer. At times, the wind blew so hard they could not work in the field, especially to put up hay.

Those incessant winds whipping around the buildings produced a mournful sound that never seemed to quit; a newspaper of those times likened this sound to that of a wolf howling in a canyon. However, this comparison did not do the wind full justice, since the wolf occasionally had to stop to take a breath while the Kansas wind never seemed to stop for anything.

In 1899, the St. Valentine's Day Blizzard, also known as the Great Arctic Outbreak of 1899 and the Great Blizzard of 1899, blasted down from Canada across the Southern United States, making it all the way to Cuba. An exceptionally severe winter weather event, it affected most of the United States, particularly east of the Rocky Mountains.

December 1898 through February 1899 was the third-coldest meteorological winter in the contiguous US, with temperatures falling as low as 35 degrees below zero in some locations on the plains. This storm had a disastrous impact across many areas of the continental US, killing crops, livestock, and over 100 people. Traffic came to a complete standstill in all parts of the country. The Crawfords were unprepared for a winter as bad as this one. Laura reported that Harmon often had to wrap his feet with newspapers when he went out to do the chores.

The Crawfords still went hunting to supplement their food supply. When it was too windy to put up hay, the whole family would get in the spring wagon with their guns and set off to hunt. Large game such as deer no longer remained on this land, but there were plenty of prairie chickens, quail, and rabbits. This fresh meat was a welcome departure from their usual diet of cured meat.

The family also ate sorghum cake, which they liked. They may have planted a little sorghum for this purpose, but since processing the sorghum was labor intensive, they probably just bought sorghum syrup at a store in Luray. Coffee and sugar were luxuries, so Candus would brown cracked wheat and boil it for coffee, and use sorghum for sugar.

Not long after the Crawfords returned from Kansas City, Mr. John Lewis, a traveling tree agent, went through the community taking orders for trees. He persuaded Harmon to plant an orchard on the hillside west of their house. Harmon enjoyed his company so much he let Mr. Lewis use the Crawford home as his headquarters, and Mr. Lewis provided the trees and grape vines for the Crawfords' orchard in exchange. The Crawford orchard included apples, cherries, grapes, peaches, and plums.

For entertainment in the late 1890s, the Crawfords and their neighbors would have dances and parties at their homes. Old and young would go—the parents would visit and play cards; the young would dance. Mail was another source of entertainment. When the snow got so deep the mail carriers could not get through, Roy would go to town on horseback with a gunnysack and bring back the mail for three or four families. The mail was a treat for the

Crawford family, who would all get around the kitchen table to read the mail by the light of their kerosene lamp. As Laura said, "We were happy in those days. We didn't know any better. Ignorance is Bliss."

10. Amherst School, with outhouse in back. (1918)

The community organized the Amherst school district on July 7, 1879, and built a sod house to serve as the first Amherst schoolhouse. This sod house soon collapsed and the community erected another. Figure 10 shows the final Amherst schoolhouse, built at the northwest corner of section 34 in the early 1890s, shortly before the Crawfords returned to the area.

At first, the school terms were only a few months long, and many of the older boys only went to school when they were not working on the farm. The early seats and desks at the school were all homemade. The children, Anna, Mirt, Roy, Bertha, Charlie, and Laura all began attending the Amherst school in the fall of 1895, and all completed their elementary education there. Numerous Bean, Beatty, O'Connor, Johnson, King, and other children also attended there over the years.

Sunday Schools had been meeting in various houses around the area or in the schoolhouse for some time, but it was not until 1887 that community members organized an Evangelical Church in the Amherst area. The Thomas Beatty family played a major role in the establishment of that Church, and the charter members were Mr. and Mrs. Thomas Beatty, Mr. and Mrs. Ben F. Bratton, Mr. and Mrs. Leander Beatty, Ettie Beatty, Zina E. Beatty, and Zenas W. Beatty.

Thomas Beatty donated approximately two acres in the northwest corner of his homestead quarter in section 34 to the Amherst Cemetery Association on June 22, 1897. This land was just south of the Amherst schoolhouse. On November 8, 1904, Mr. Beatty donated additional section 34 land to the Evangelical Association of America, bringing the total for the church, cemetery, and schoolhouse to 4.114 acres.

After completion of construction in 1906, the church building on this land became the center of community activities. A full basement dug under the church in 1923, provided room for Sunday School and fellowship occasions.

Church membership reached a maximum of 97 with an average attendance of 75 early in the 1900s. Membership started to diminish by the early 1920s as the farms became larger and there were fewer people living in the community. The Crawford family, although living nearby, was never closely involved with this church, but they did participate in some of the community social activities held at the church.

Progress and Prosperity

By the late nineteenth century many ingenious mechanical designs had led to the invention of implements that mechanized most of the farm tasks and revolutionized farming. Harmon gradually started to purchase a few of these new farming implements. He was cautious about this, refusing to go into debt and only buying what he could afford to pay for with cash.

One of the items high on his priority list was a horse-drawn sickle-bar mower so that he would no longer have to cut the hay or the grain crops with a scythe. The mower utilized a reciprocating sickle that moved horizontally across some fixed sharpened teeth to cut the stalks of the grass or grain. Machines of this type, invented in the mid-1800s and readily available (about $40) in the 1890s, quickly took over the tasks of cutting hay or ripened grain, replacing the backbreaking work previously done with a scythe. Harmon also purchased a horse-drawn dump rake ($20) to gather the dry hay from the ground and deposit it in a windrow. Roy and Charlie helped Harmon load the hay from the windrows onto wagons, using pitchforks for this task.

Somewhat later, Harmon purchased a reaper-binder ($125). This implement cut the grain using the same sickle-bar mechanism as did the mower, and it bound the cut grain stalks into bundles tied with twine.

Harmon bought a McSherry grain drill (about $40) to plant his grain more rapidly and more uniformly. The McSherry drill, like other grain drills available in the 1890s, made a marked improvement in Harmon's planting of wheat and other small grains. A seed box atop the drill held the seeds not yet planted. Eight runners glided along under the surface of the soil, slicing open a series of equally spaced uniform depth furrows. Geared mechanisms then dropped seeds a pre-selected distance apart in each of these furrows and metal disks pressed the sides of each furrow down over the seeds just firmly enough to provide the proper conditions for germination and growth. This drill greatly sped up the planting process, made more efficient use of the seed grain, and resulted in crops more uniformly distributed in the field and thus able to produce higher yields.

By the turn of the century, most farming in central Kansas had shifted from mule-power to horsepower, largely because workhorses were nearly twice as powerful as mules. Another advantage was that horses could produce offspring while mules, being hybrids, could not. The Crawford farm was no exception. The greater degree of mechanization of the farm operations meant that the farm needed several workhorses and all the associated harnesses and paraphernalia. Harmon did keep a pair of mules to pull the wagon, since mules were more sure-footed than horses and were better over rough terrain.

These horses and mules required a pasture and an adequate supply of water. Pumping enough water from the well to a stock tank to satisfy all these horses and other livestock had become a huge job, so Harmon purchased and installed a windmill ($50) to pump water from their well to the tank. He also built a pond in the pasture south of the farmstead. That pond supplied water for the horses, mules, and cattle during parts of the year, but it usually went dry during other parts of the year, necessitating heavy use of the stock tank at those times.

More horses also meant an ever-increasing supply of horse manure that Harmon could use as fertilizer on his fields. Farmers had to spread the dung by hand until Joseph Kemp invented the first successful mechanical manure spreader in 1875. This was a ground-driven unit pulled by one or more horses or mules. By the early 1900s, Harmon had purchased one of these manure spreaders ($16).

In the late 1800s, threshing the grain required hand or horse power, and was still a labor-intensive task. Fortunately, rudimentary threshing machines began to appear about 1890, and steam engines started to replace horses and mules as the power source for these machines. Threshing machines powered by steam engines were particularly effective, greatly reducing the amount of labor required.

Soon many of the farmers were having their grain threshed by these huge machines. This made it practical to plant more land to wheat, and Kansas farmers planted more than five million acres to wheat in 1900 and were planting even more with each succeeding year. However, these threshing machines were expensive, about $1,200-$1,300 for the threshing machine plus the steam engine to operate it, too expensive for each farmer to own one. This gave rise to custom threshing businesses with the owners of the threshing machines threshing grain for other farmers for a fee. It typically cost a few cents a bushel to thresh wheat that eventually sold for about eighty cents a bushel. Harmon was hiring custom threshers to thresh his grain by the early 1900s.

Each of these new implements was a mechanical marvel, but none was as complex or generated as much exhilaration as did the threshing machine. Threshing was a time of excitement and of promise; a dirty, noisy, and labor-intensive process, requiring skill and teamwork from a large threshing crew.

Horse-drawn wagons hauled the bundles of grain from their shocks to the threshing machine, where crew members used pitchforks to pitch them into a feeder or hopper. A rapidly rotating set of blades called the separator tore the bundles apart and then beat the straw and heads onto a grooved plate, knocking the grain from the heads. A straw rack and a series of progressively smaller shaking screens removed most of the remaining straw and the chaff from the grain, and a stream of air blew the remaining chaff and straw away. A stronger blower blew the straw and chaff onto a straw stack, and the cleaned grain then fell into a hopper, where an auger elevated it to a bin.

While the threshing was ongoing, Candus and the Crawford daughters had to start early in the morning to prepare the massive amounts of food to feed this large hard-working and hungry crew. If necessary, they could call on some of the neighbor women to help them with this task. That golden stream of grain emerging from the spout of the threshing machine represented much of the year's income for the Crawford family, so all anxiously awaited a successful completion of the threshing.

Horses powered all these farming implements except for the steam-powered threshing machine. With more horses and that new machinery, the amount of labor required to produce wheat or hay was less than half what it had been on the Crawford farm in the 1880s. By the mid-1890s, Roy was getting old enough to operate a team of horses or mules as they pulled some of the farm implements.

With the new implements and Roy's help, Harmon was able to farm more land. He still refused to go into debt for anything, including land, but by the early 1900s he had accumulated enough cash to purchase the northwest quarter of section 21 (NW¼ 21-11S-13W) from John and Mary Marschang on April 7, 1904 for $3,200. He also purchased the southwest quarter of section 35 (SW¼ 35-11S-13W) from the Union Pacific Land Company on May 19, 1904 for $1,358.

By 1900, Luray had two grain elevators, a lumberyard, and a small stockyard, all located adjacent to the railroad tracks. The town also had a school, churches, and at least 24 businesses including a bank, which together occupied both sides of Main Street for nearly a block and a half.

11. The Pospishil Opera House in the early 1900s when it housed the Fallis & Grammon general merchandise store. Judging by the crowd, this store was quite popular.

Some of these businesses were in limestone buildings, while wooden-frame buildings served for others. Most of these were one-story buildings, although some included attic storage space. Luray also had the Pospishil Opera House, a two-story limestone building constructed in 1898, where traveling groups could provide concerts, variety shows, and lectures.

One example was M.C. Lampke's Moving Picture Show, presented on December 20, 1902. As described in the *Luray Herald*, this presentation utilized an Edison Kinetoscope to show pictures of the Philippines and of battleships from the Spanish-American War, while a scout who saw service in that war described the action.

However, the Pospishil building only served as a venue for entertainment for short period, and within a few years the Fallis and Grammon general store (figure 11) occupied that building instead. Main Street was still a dirt street, but wooden or concrete sidewalks fronted most of the buildings (figure 12).

12. South end of Luray Main Street, early 1900s. Fourth of July parade.

Waldo was also a thriving town by 1900, with a small stockyard and several grain elevators operating next to the railroad tracks. The town also included a lumberyard that supplied building materials for the community, and at least 16 businesses, including a millinery shop and a hotel, that lined a block and a half of Main Street on both sides. Investors organized the Waldo State Bank in 1902 with a capital of $6,500.

As the population of the Amherst community increased so did the need for roads to connect the settlers with one another and with the towns of Luray and Waldo. Various acts of the Kansas territorial legislature and later of the Kansas state legislature declared that all sectional lines in most Kansas counties "be and hereby are declared public highways." This gave the counties the right-of-way to open actual roads along section lines at their discretion. The late 1880s and the 1890s were periods of active road building in the area, and by 1900 an extensive set of county roads followed section lines across the Amherst community and the areas between there and Luray and Waldo, and connected to Russell and Bunker Hill. These roads eliminated the need for the old stage coach routes, and those quickly fell into disuse.

In the early 1900s Roy, Bertha, Charlie, and Laura were still living at home. The two oldest Crawford daughters, Anna and Mirt, had married and moved out of the Crawford household. Mirt Crawford married Ralph Thomas on October 6, 1898, and they were living with Ralph's parents south of Bunker Hill. Anna

Crawford married Frank Caprez on October 31, 1900, and they lived on a farm near Waldo. Mirt and Ralph had no children, but Anna and Frank had four: Viola, Amos, Ruth, and Frank Raymond. Bertha Crawford married Commodore Perry "Dode" Cochrun on October 27, 1904. They initially lived in a place they rented in Luray, but later moved to a farm they rented near Waldo, where they raised their three children: Virginia, Doris, and Delbert.

13. Harmon and Candus Crawford. (1898)

Roy Crawford, my grandfather, was now in his early twenties and was ready to branch out on his own as well. Each year he followed the corn harvest north into Nebraska and the Dakotas, shelling corn during the winter to make extra money. He used some of this money for a down payment on the northwest quarter of section 35 (NW¼ 35-11S-13W), which he purchased from John Pospishil on September 1, 1904 for $2,720.00.

This land, adjacent to the Harmon Crawford homestead quarter, was originally railroad land, and Pospishil had purchased it from the Union Pacific Railway Company on April 15, 1897. Mr. Pospishil had not yet broken the ground there or built on it by the time he sold it to Roy.

To buy this property, Roy took out a mortgage of $1,220 from Mr. Pospishil, which Roy soon repaid as he continued to make these trips north to shell corn. This apparently was the only time Roy ever went into debt.

14. Harmon and Candus Crawford homestead. The stone quarry Harmon used shows up as a light area on the hillside to the right of the stone house. (1908)

By 1906 the Beatty families owned three quarters of section 34. The Thomas Beatty farmstead in the northwest quarter had

71

become substantial with an extensive set of outbuildings. Zenas Beatty, one of Thomas Beatty's sons, had owned the southeast quarter of section 34 since 1898, and was living on that land, and Zina Beatty, another of Thomas Beatty's sons, purchased the southwest quarter from the Agricultural Land Company on June 15, 1906. Harmon Crawford still owned the northeast quarter of section 34, and his farmstead contained several outbuildings as well as the stone house he had built (figure 14).

By the end of the nineteenth century, Harmon and Candus were living comfortably in their expanded stone house on the farm they had homesteaded and now owned. Their six surviving children had grown up on this farm and were approaching or had already reached adulthood. The farm was prosperous and had a set of outbuildings appropriate for all the horses and other animals. An adequate water supply supported the livestock.

They had survived and prospered despite the many obstacles Mother Nature had brought their way—blizzards, droughts, dust storms, and the never-ending wind, and they had done so without ever having to go into debt. They had witnessed and taken advantage of the technological advances that had occurred rapidly in the late nineteenth century, so that now they required much less labor to produce their crops. This enabled them to increase the size of their farm, and they had achieved a level of income and savings that allowed them to purchase more land, setting the stage for future expansion.

They had indeed transformed this piece of the virgin Kansas prairie into a viable farming operation that they could pass on to the next generation.

Second Generation

PROSPERING THROUGH DIFFICULT TIMES

Turning Over the Reins

With some trepidation, Leopold Hampl, along with his wife Marie (Triner) Hampl and their nine children, stepped off the ship *Berlin* and onto American soil in 1873. They had just arrived from Europe where they had left their native Bohemia to create a new life for themselves in America. The Hampls came from Tytry in Bohemia, where Leopold had owned a farm and a mill. They were relatively well off financially, so they did not immigrate to America to better their family economically. Instead, by 1873 the two oldest Hampl sons, Raymond and Alexander, were reaching the age when they would be subject to conscription into the Austrian army, and this provided the incentive for the family to leave Bohemia. The Hampl family went immediately by train from Baltimore to Milwaukee, where they spent the winter. In the spring they traveled by train to West Point Nebraska, where many other immigrant families from Bohemia had already settled.

In West Point, Alexander "Alex" Hampl met Agnes Janecek, whose family had emigrated from Bohemia to America and settled in West Point in 1872. Alex married Agnes in 1879, and they started their life together on a farm in Saint Charles Precinct in Cuming County, Nebraska. In 1886, Alex, Agnes, and their three children moved to Kansas, where Alex bought a quarter section of land (SW¼ 1-12S-13W) from the Union Pacific Railroad Company in Russell County for $660. This was on the eastern edge of the Amherst community.

While in Kansas the Hampls suffered through the terrible blizzard in the winter of 1886, and this convinced Alex and Agnes to return to Nebraska for a while. Their daughter Albina, my grandmother, was born in Nebraska on December 16, 1887.

Ten years and several children later, Alex and Agnes and their family moved back to that Kansas land, traveling in a covered wagon, and cooking in it until they built a house to live in. As soon as their wagon came to a stop on their Kansas land, four-year-old Philip, their son who was anxious to get back on the ground, jumped out of their wagon. Unfortunately, he was barefoot and landed in a patch of sandburs, a rude and unexpected welcome to their new home.

The Hampls found that piece of land to be poor farmland, and they tried several other sites for their farm before finally settling on a different quarter section (SE¼ 36-11S-13W) in the Amherst area. They bought this quarter from A. J. and Jennie Glaze in 1908 for $5,000. There was an almost-new frame house on this land, so the family was able to settle into that house immediately. By the time

78

they arrived at that site, Alex and Agnes had ten children: Jim, Agnes, Tom, Albina, Philip, Mary, Anna, Rose, Alex Jr., and Bill. The family liked that house and farm, which was less than three miles from the Crawford farm, and they enjoyed their life there.

15. Agnes and Alex Hampl. (ca. 1880)

In addition to being a farmer, Alex was a good blacksmith and had been an apprentice file-maker before coming to America. He could take a good piece of steel and make a cold chisel or drill bit, temper it, and then sharpen it. He made many of his own tools. Once the Hampl sons were old enough to help with the field work, the household developed and managed a sizeable farm operation, as was necessary to support their large family.

In an interview recorded in 1980, Bill Hampl reminisced that the Hampls usually managed to eat well. They would butcher a couple of hogs every fall. They didn't have an icebox or refrigerator in those days, so after they butchered, Agnes would fry meat for several days. Then she poured lard over it to seal it. Alex built a smokehouse for the hams and the bacon, which he would smoke for about a week. When he took them out of the smokehouse, the family would wrap them with cloth and sew it tightly to keep out the dirt. After that, they would hang the bacon and hams up in the granary until it was time to eat them. Bill said that all this "made pretty good eating," but he also said that some of this preserved meat lasted until it "got kinda strong in late summer." The Hampls also frequently ate chicken, often with Bohemian dumplings called knedliky.

On Saturdays, the Hampl children would go out in the cornfields and bring in corn stalks. One of them would feed the stalks into the stove for fuel to keep it hot while their mother made kolaches, yeast dough rounds topped with a sweet mixture. Favorite Hampl toppings included prunes, apricots, and poppy seed. On Sunday morning, they would have kolaches and coffee for breakfast.

Apples were also an important part of their diet. The family would buy ten to fifteen bushels of apples at a time. Alex would pick out the bruised or badly spotted ones and put the rest of the apples in boxes to save until later. The kids were to eat the bruised ones first. Every so often Alex would go through the apples, pick them over, and put the better ones back in the box, leaving the rest for the kids. Bill and Alex Jr. recalled that they were "eating only the rotten apples all the time!"

Harmon Crawford died of a heart attack on May 30, 1908 at the age of 58. He was buried in the Luray Cemetery. His death led to a redistribution of the land he had owned—Harmon died intestate, so Candus inherited one-half of his estate and the six children each inherited one-twelfth of the estate. After a complex series of swaps and buyouts, by 1913 Charlie owned the Crawford quarter section of farmland in section 21 as his farm, Candus had gained sole ownership of the homestead quarter in section 34, and Roy owned the southwest quarter of section 35 to go with the northwest quarter of that section that he had previously purchased. The Crawford daughters had sold their shares to Roy, Charlie, and Candus, so they ended up with money instead of land.

The year 1908 was eventful for another reason; Roy Crawford married Albina Hampl on October 28, 1908, in Russell, Kansas. With Harmon gone and with Roy now married, my grandparents Roy and Albina became the *de facto* leaders of the Crawford farming activities in sections 34 and 35.

16. Roy and Albina Crawford. (1908)

Roy was a resourceful person who had already purchased property of his own by the time he and Albina were married. A man of integrity and an astute businessman, he had grown up learning farming from his father. Somewhere along the way he had also picked up important skills as a mechanic, a machinist, and a blacksmith, all of which he used in support of his farming.

Albina (Hampl) Crawford was a superb cook, seamstress, and housekeeper as well as having the other skills that came from growing up in a large Kansas farm family. Most of the Hampl family was musically talented, and Albina was no exception although she did not exhibit her talents publicly. She could play the piano, and was always humming different songs as she went about her work, occasionally singing a few words before resuming the humming.

17. Candus' new frame house. Opening the many exterior doors improved the air circulation during the hot Kansas summers; the upstairs sleeping porch provided a cool, bug free, place to sleep during periods of extreme heat. (ca. 1910)

Not long after Roy and Albina were married, Candus, Charlie, and Laura went to Pennsylvania for the winter, leaving Roy and Albina to live undisturbed in Candus's stone house until Candus and the others returned in March 1909. Once she returned, Candus supervised the building of a frame house near the site of the stone house, and by the end of the summer of 1909, Candus, Charlie, and Laura were able to move into that frame house, allowing Roy and Albina once again to have the whole stone house to themselves.

Candus bought a Model T Ford in 1910, but since she didn't drive herself, Charlie drove her wherever she needed to go, and this car in effect became Charlie's.

Laura and Charlie were soon making their separate ways. Laura Crawford married John Luder in March 1913, and they lived in Waldo where John worked as a carpenter and mechanic. They had one child: Mabel, born in 1914. Charlie Crawford married Minnie Carter in July 1916, and they lived in Candus's house until

18. Candus' new Model T (Charlie's car). Charlie is in the car and Candus is standing beside it with her granddaughter Virginia Cochrun (Bertha's daughter). (1910)

they completed building a house of their own on Charlie's land in section 21. They had a daughter Carol Jean, born in 1932.

19. Barn, well with windmill and stock tank, stone house, and Candus' new frame house at the Crawford homestead. Note the spring wagon by the well and the stone post fences. The foreground shows ruts in the muddy road running past the homestead. (1919)

After Charlie's marriage in 1916, Candus moved into a house in Luray where she was living by herself until she married her second husband, Jonathan "Jack" Norris on October 13, 1921. After that, she lived with him in Luray until he died in April 1928. Candus continued living in Luray, and died June 10, 1933, at age 85. She is buried beside Harmon in the Luray Cemetery.

By the 1920s, all the Hampl daughters had married and left home. Among the sons, Bill was working as a farm hand for Roy Crawford, Jim was working as a farm hand elsewhere, and Tom, Philip, and Alex Jr. were married and were farming on their own. Tom had married Anna Radina in 1911, Philip married Mathilda Bachmann in 1918, and Alex Jr. married Sarah Griffin in 1920.

Except for Jim, all the Hampl sons were living within about three miles of the Crawfords.

Alex and Agnes Hampl retired from farming in 1920, and they moved to Luray. Delmar Hampl (Philip's son and Albina's nephew) remembered their Luray house and noted, "We visited often when I was about 5-6 years old. Mom, Grandma, and I would sit in the living room with very little conversation because Grandma Hampl knew very little English. My Dad and Granddad Hampl would go to the cellar and pull a couple bottles of beer from a little well in the cellar and sit and visit."

Another significant change in land ownership on section 34 resulted from the death of Thomas Beatty in August 1909, at the age of 79. His wife Jane continued to live on their homestead until her death in 1915. Their son, Zina Beatty, his wife Ora, and their children also remained there until 1912, when they moved to Colorado Springs.

After Jane's death, the Beatty homestead quarter section in section 34, minus the roughly four acres Thomas had previously donated to the Amherst church, passed by will to Zina. In 1920, Zina sold this quarter section to Russell M. Cochrun. Russell, his wife Alma (Bratton), and their four children Freda, Eloise, Kenneth, and Eunice lived a mile east of Roy and Albina's house. Roy Crawford would eventually purchase this quarter section from the Cochruns.

Roy and Albina Settle In

Soon after his marriage to Albina, Roy began to establish his own farmstead, constructing a frame house and outbuildings on a site he had chosen on his northwest quarter of section 35. While he was building that house, he and Albina continued to live in the stone house on the Crawford homestead quarter in section 34, and their daughter Pauline Violet Crawford was born in that stone house on May 14, 1913. Later in 1913, they moved into their new house on section 35, and their son Clarence Richard Crawford, my father, was born there on July 9, 1915.

Roy and Albina designed the house built on their farmstead. Roy did much of the construction work himself, getting help from some of the neighbors and from a skilled carpenter when necessary. It was a frame house, somewhat smaller than the frame house that Candus had built on her farmstead, and it had a family room on the first floor where they ate their meals and where they relaxed once they had finished all the chores. The first floor also contained a kitchen, a guest bedroom that served as Albina's sewing room, and a parlor that the family used when there were guests or when someone wanted to play the piano located there. A coal-burning stove in the middle of the family room provided the only source of heating for the house in the winter.

On the second floor, there were three small bedrooms, along with an exterior sleeping deck that provided a cooler spot for sleeping on hot summer nights.

Roy fenced in a large space in his new farmyard for a garden, and this garden became Albina's passion. She filled this garden with a wide range of vegetables and herbs for her cooking, and carefully nursed a variety of flowers there.

20. Roy Crawford house. This house had only four exterior doors, fewer than in Candus's house, but by opening those four doors there was good air circulation during the hot Kansas summers. This house also had an upstairs sleeping porch to provide a cool place to sleep.
(1922)

The kitchen range they used for cooking burned kerosene (fuel oil). Behind the house, there was a cistern and another small building where Albina did the laundry. Figure 20 shows the elaborate roof guttering used to collect rainwater from the roof and direct it to the cistern. Except in extreme droughts, the cistern provided all the water they needed for drinking, cooking, weekly baths, and laundry. A hand-cranked bucket pump raised the water from the cistern and dumped it into a pail or other container, which they carried to wherever they needed it.

There was no running water or indoor bathroom, but the family did have an outhouse. Chamber pots in the upstairs bedrooms served when nature called in the middle of the night; it was someone's chore to empty them outside the next morning.

By the mid to late 1920s, radios had become popular and common. The Crawfords installed a wind-charger (a windmill-driven generator) to provide electricity to charge a set of storage batteries in the cellar. Those batteries powered their radio. Roy also wired light fixtures in the house, but Roy and Albina almost never turned on the lights, preferring instead to use a kerosene lamp for light or else to sit in the dusk and the dark until it was time to go to bed.

In addition to their frame house, their farmstead included a well and a barn and outbuildings. There are no photos of the farmstead at that time, but pictures taken in the 1920s show that by then it had become an extensive set of farm buildings surrounded by a shelterbelt of mature trees.

21. Roy Crawford farmstead, viewed from the southwest. Note house in the distance, barn in center foreground, windmill at well in right foreground, and stone post fencing. The stand of trees (shelterbelt) encloses the house, barn, and yard. (1924)

22. Roy Crawford farmstead, viewed from the top of the windmill at the well. (1924)

Roy designed the barn on their farm to his unique specifications. It was a substantial building, and included a row of pigeon nests under the eaves. These nests were open to the outside, and with access from the inside so that Albina could catch the young pigeons (squabs) which she cooked in cream—a delicacy.

The barn had an extension of the peak of the roof (figure 22) to facilitate loading the hay into the hayloft.

Putting hay into the hayloft was an elaborate operation. The extension of the barn roof accommodated a pulley, and a rope ran up through that pulley and back down to a device that picked up the hay. A team of horses driven by Albina pulled the other end of the rope, thus lifting the hay to where the men in the hayloft could pull it into the barn through the upper door. Albina then had to back the horses to lower the device for another load of hay, a difficult process that she hated.

Roy dug the well on their farmstead into the layer of shale below the soil, so the water that seeped into it was potable, but there wasn't a lot of it. Roy wanted to have more livestock in that quarter, but the well only supplied enough water to support one team of horses and one cow in the area near the barn. On March 27, 1916, he purchased the southeast quarter of section 35 (SE¼ 35-11S-13W) from Marcus L. Bratton for $6,300. This gave him three contiguous quarters in section 35. A well in that southeast quarter provided a supply of good water adequate for the rest of his work horses, which he then kept in a pasture near that well.

Roy also had a shed he used for his shop. Later, he restricted the shop to half of that shed so he could use the other half as a garage for their car. This shop contained a blacksmith's forge that burned coal and had a hand-cranked blower to provide a forced draft to make it burn hotter. Nearby was a large blacksmith's anvil where he beat the heated iron or steel implement into the desired shape.

He probably used the forge and anvil to sharpen plow shares by hammering out the red-hot cutting edge. That edge frequently had to be resharpened at the correct angle to keep the plow in the hard soil. The forge burned anthracite coal rather than the less expensive bituminous coal to produce a hotter flame for softening the metal pieces. A big vise anchored to the workbench held items under construction when necessary. A hand-cranked grinding wheel attached to the workbench served to sharpen blades or to shape other metal items.

This shop remained largely unchanged into the 1940s and 1950s. It fascinated me as a child and I was thrilled to watch the sparks flying when Grandpa used the grinding wheel to sharpen the blade on the hoe. I was even more excited when he would start a coal fire in the forge and allow me to turn the crank for the blower to make the fire glow brightly. The wisps of smoke from the embers mingled with the faint oily smell of the garage, and the whirr of the blower coupled with the sounds of the air rushing through the coals added to the ambiance, making it into a real adventure.

When I was a little older, I would heat a piece of steel in that forge, and then put the hot steel on the anvil and beat it into a new shape. The smell of the glowing hot steel and the sound and feel of my hammer striking the steel on the anvil made me feel a bit like a real blacksmith. That was fun—but although I sometimes got as dirty as a real blacksmith there, I never acquired any real blacksmithing skills and I never managed to make anything useful with the forge, hammer, and anvil.

Both Roy and Albina loved to travel, delighting in any chance to see new sights, or even to revisit familiar areas. In 1913, shortly before Pauline was born, they took the train to the west coast to visit Roy's sister Mirt (Crawford) Thomas and her husband Ralph, who were living in Portland Oregon.

Along the way, they left the train in Syracuse in western Kansas to visit Albina's sister Agnes (Hampl) Beach and her family. The Beach family was away for the day, and when they returned home, they saw that a light was on, and discovered that Roy and Albina were there. As Amy Beach remembered, Roy and Albina had hired a man with a lumber wagon to take them the 18 miles from Syracuse to the Beach home, and they were already cooking a couple of ducks they had brought to make dinner for both families. The Crawfords spent several days with the Beaches before returning to Syracuse to catch the train to continue their trip west.

By 1915, wheat had become the primary crop for our farm, as well as for many of the other farmers in western Kansas. Drought and a winter freeze during 1916-1917 led to winter kill of the wheat, destroying about 60 percent of the crop planted for the 1917 harvest. That drought continued during the summer of 1917, and the scarcity and high price of seed wheat limited the acreage planted to wheat for the 1918 crop. It also led to later than normal planting times, resulting in winter kill of some of that crop as well.

Although World War I in Europe had created a high demand for wheat, pushing the average price up to about $2.00 per bushel or higher, most farmers, including the Crawfords, didn't have enough wheat to sell to provide them with adequate income for those years.

However, the War had created a demand for skilled workers, and since Roy was mechanically skilled, he was able to find a wintertime job in a machine shop in Rocky Ford, Colorado, to supplement the family's income. Roy, Albina, Pauline, and Clarence spent the winter of 1918-1919 in Rocky Ford while Roy was working there.

In 1918 or 1919, Roy and Albina bought a Ford Model T Touring automobile ($525). Roy and Albina had a very fine horse and buggy, but this may have been the first automobile they owned. According to Pauline, Albina didn't like to drive and was not good at driving. Therefore, Roy did nearly all the driving until Pauline and Clarence were old enough to operate the car themselves. Figure 23 shows this car, and figure 24 shows that this vehicle was sufficiently robust to ford the Saline River, still an important capability in those days. Apparently, it could also handle the snow and cold weather and did not have to suffer the indignity of requiring a team of horses pull it, as did Charlie's car in figure 25.

23. Clarence and Pauline fill the radiator of the Crawfords' new car. (1920)

24. Fording the Saline River. (1920)

25. Three modes of transportation through the snow; riding a horse, horses towing your car, or motoring along in a car. (November, 1925)

Cars offered new entertainment opportunities as well as becoming important to everyday activities. Bill Hampl recalled that when the Hampl boys were older, every so often they would have to have a race. Alex Jr. bought a second-hand Chevy and their dad had a Model T Ford. On Saturday or Sunday, they would go out south of their place and race down that hill. According to Bill, the old Model T didn't run very fast, "Fifty miles an hour would be the top speed you could get out of them — well, downhill they'd go pretty fast."

Early in the 1920s, Albina developed respiratory difficulties and the doctor recommended that she spend some time in a warm dry climate. This led to an extraordinary ten-month trip to the west coast that the family took in 1922-1923, when Pauline and Clarence were nine and seven, respectively. After extensive preparation, including building a "RV" (a cabin built on a Model T truck chassis) and equipping it for travel, Roy, Albina, Pauline, and Clarence began their westward trek in the late summer of 1922. They had a new camera, and took hundreds of photos while on this adventure. Those pictures still exist and many of them have been published in a book about that extensive trek.

26. Roy, Pauline, Albina, and Clarence Crawford. (1919)

Their trip took them to Colorado, Wyoming and Yellowstone Park, and Oregon before turning south into California. On New Year's Day, 1923, Roy, Albina, Pauline, and Clarence saw the Rose

Parade and the floats in Pasadena and Roy attended the first Rose Bowl football game held in the new Rose Bowl Stadium. The family went on to the San Diego area, where they settled at a municipal campground for several months, enjoying the warm weather there and visiting the nearby attractions.

In the spring of 1923, the Crawfords headed home, stopping at the Grand Canyon, the Petrified Forest, and at Santa Fe, New Mexico along the way. They were home on the farm by early June, in time for the 1923 wheat harvest. Bill Hampl looked after the Crawford farm, tending the livestock and the crops during the time they were away. Albina was much healthier by the time of their return, so the trip had served its intended purpose in addition to providing a remarkable experience for the entire family.

On that 1922 trip west, they passed through the wheat-growing area on the Columbia Plateau in eastern Oregon. Stopping to visit with farmers harvesting wheat there, Roy discovered that eastern Oregon had an even lower annual rainfall than did central Kansas, but that the Oregon farmers were still able to raise good wheat crops by summer fallowing their land. This meant that every other or every third year they would not plant wheat on some of their land, letting that part of their land lie fallow throughout a normal growing season to allow replenishing the moisture in the soil, The farmers still had to do sufficient tilling of the land to prevent the weeds from using up all the moisture, but otherwise that piece of land would lie idle for the year. The next year they would plant a crop on that land and leave a different piece of their land fallow.

Roy thought this was such a good idea that he decided to try summer fallowing on his land. In the spring, he completed most of the pre-planting tilling operations on the land that had been lying fallow. After harvesting the current crop, he left the straw stand on the land and would not till it until the fall when he was ready to plant the new wheat crop.

Delmar Hampl noted that, in August, while all the other farmers were trying to carry out the pre-planting tillage, Roy was free to take a vacation and go fly-fishing in Colorado. He was one of the first in the Amherst area to employ the summer fallow technique, and his land was producing 20 bushels per acre while others were getting only 13 bushels per acre on their continuously cropped land. Roy could do this because he owned his land and could decide what crops to plant. Farmers renting their land had to plant whatever the landlord required, and landlords often required planting all the land. However, Roy's summer fallow results were so good that summer fallowing eventually became a standard practice in the area.

Community and Schools

In the 1920s and 1930s most of the extended family lived nearby and family gatherings were common; some were for special occasions but others were just because the family felt like it. The Crawfords and the Hampls both had many such gatherings in which Roy, Albina, and family participated. A few of these get-togethers warranted picture taking. Figures 27-29 provide examples.

27. Hampl picnic at Shady Bend, about 1925. **Back Row**: Alex Hampl, Agnes Hampl, Tom Hampl, Alex Hampl Jr., Harvey Bean, Mrs. Robert Chudomelka, Philip Hampl, Robert Chudomelka. **Middle Row**: Anna Hampl (Mrs. Tom), Albina Crawford (Mrs. Roy), Sarah Hampl (Mrs. Alex Jr.), Mary Bean (Mrs. Harvey), Mathilda Hampl (Mrs. Philip). **Front Row**: Leonard Bean, a Chudomelka, Clara Hampl, Leona Bean, another Chudomelka, Pauline Crawford, Irene Hampl, Clarence Crawford, Frank Hampl, Elmer Hampl.

28. Extended Crawford Family. **Back row**: Charlie Crawford, Dayton Thomas, Minnie (Carter) Crawford, Albina (Hampl) Crawford, Commodore Perry Cochrun, Virginia (Cochrun) Thomas. **Middle row:** Mabel Luder, Pauline Crawford, Doris Cochrun, Candus Crawford, Bertha Cochrun, Laura (Crawford) Luder, John Luder. **Front row**: Delbert Cochrun, Clarence Crawford. (1928)

29. Crawford Siblings: Charles, Laura, Bertha, Anna, Mirt, and Roy, at Roy Crawford house. (1934)

In addition to family gatherings and occasional community potluck suppers, considerable socializing among the rural community members occurred during drop-in visits. Once automobiles were generally available, it became common for folks to hop into the family car in the evening, drive over to one of the farms in the neighborhood, and drop in unannounced, expecting to spend the evening visiting. These visits might be with relatives or with unrelated neighbors. Most folks considered this normal neighborly behavior, and enjoyed the visit as a break from the usual routine. In the Amherst community, this practice continued to be a significant part of the social life of the community at least through the 1950s.

30. Fishing at Shady Bend. (1924)

Fishing and hunting could be recreational and social activities, and they supplied welcome variety to the diet. Roy loved to fish, but he claimed he was allergic to cleaning the fish so Albina always was stuck with that task. Roy did not seem to be allergic to eating the fish though.

Alex Hampl Jr. also liked to fish, and he and Roy were competitive friends, so they kept their favorite fishing holes secret from one another. Two favorite places were the Sand Rocks, an attractive rock cliff along the Saline River a few miles south of the Crawford and Hampl farms; and Shady Bend, the site of a mill dam in the Saline River east of Lincoln, Kansas.

31. A successful hunt. Roy is at the left. (1925)

Hunting was also an important activity, providing recreation and a source of food. The standard uniform Roy wore for both work and leisure was his signature blue bib overalls over a white

103

or light-colored long-sleeve shirt, and he wore a necktie with this uniform even for work. Figure 31 shows the results of a successful hunt, with Roy wearing his standard garb.

Saturday nights in town were an anticipated social occasion, the time that folks would congregate in the towns to do their shopping and visit with friends. Children would run around playing hide and seek or other games. Teens might gather at the soda fountain in the drug store. Old men would congregate in the pool hall to play cards, or stand in front of one of the stores swapping lies with one another. A bandstand on Main Street hosted numerous planned or impromptu musical performances, and at frequent dances, young couples danced to bands formed by local musicians.

Throughout the 1920s, Luray remained vibrant as the commercial center for a large farming community. Pangburn's department store, now occupying the former Pospishil Opera House building, provided a variety of merchandise to meet the household needs of their customers. The First National Bank, grocery stores, grain elevators, lumberyard, hardware store, the *Luray Herald* newspaper, and a few other businesses provided most of the other goods and services required by the community. Several churches, stone grade school and high school buildings, a telegraph and express office, and a post office having two rural routes satisfied religious, educational, and communication needs.

Waldo was also thriving in those days. Patrons could withdraw money from the bank and dine out in a choice of three different cafes, while leaving their horses stabled at either of two livery stables. There was a lumber yard, and three different stores to meet hardware needs. Three general stores, a meat market, a furniture store, two coal dealers, and several icehouses provided options for food, clothing, fuel, and household comforts.

Two automobile dealers, three automobile repair shops, a harness shop, three grain elevators, several produce buyers, and a farm machinery dealer were available to provide for farming and transportation needs. There were also several doctors and dentists, two drug stores, two barber shops, and a watch repair and jewelry shop. The *Waldo Advocate* was still the weekly newspaper for the community. The town had also found it necessary to provide a jail for those who occasionally "took their drinking too seriously," likely at a party or at the local beer hall.

Four daily passenger trains—two eastbound, and two westbound—provided passenger services to Luray and Waldo. In addition, one eastbound and one westbound regularly scheduled freight train per day delivered a variety of freight to and from those towns. The passenger train service, together with the rapid rise in the number of automobiles, ushered in a significant increase in the mobility of the rural population, and the freight service provided the lifeblood for the commercial activities thriving in the towns and surrounding communities.

The freight trains played a crucial role for the farmers during harvest time. Each of the grain elevators at Luray and Waldo had some grain storage capacity, but their combined total storage capacity was far less than the amount of grain harvested in that area in any given year. Once their grain bins were full, the elevators could no longer purchase the grain from the farmers, as they had no place to store it.

The elevators depended on the railroads to supply them with freight cars that they could load with grain from their storage bins, freeing up storage room so they could purchase more grain. The railroad would then transport the full freight cars to the large central grain handling facilities. Whenever the elevators received insufficient freight cars to handle grain at the rate harvested, the net result was that the farmers had to find some place to store the grain until the elevators once again had storage room. Roy had a few grain bins built into his barn; some of these held oats for the horses, but others stored wheat when necessary.

The farmers didn't receive income from the harvest until they sold the grain to the elevators. Since large portions of the country harvested wheat at about the same time, there was usually a shortage of appropriate railroad cars around harvest time, and this created severe cash-flow problems for the farmers. News that one of the elevators had received a railroad car to fill precipitated a rush of farmers hauling loads of grain to the elevator, trying to get there while the elevator still had space.

The childhood experiences of Pauline and Clarence were not unlike those of many other children growing up on farms, where they enjoyed the freedom and relative independence that the farm environment offered. Of course, they had age-appropriate chores to do. Pauline recalled that when she was quite young, she walked the half mile to the King's farm to buy cream. She carried the cream back home to churn it into butter.

Pauline and Clarence had relatively good luck with their health, avoiding the serious and often deadly childhood diseases. However, like most children, they suffered through numerous colds and other routine childhood ailments, although their remedies were sometimes a little different from now. According to Pauline, a common cold remedy consisted of sugar mixed with whiskey. She never did see the whiskey bottle anywhere at any other time. She may have spent some time looking for it, though, since she thought that the "whiskey and sugar tasted pretty good."

At age five, Pauline started to school at the Amherst school in the fall of 1918. This was the same school and the same building where Roy and his siblings had attended years earlier. She recalled that "I got into trouble my first day at school. The school had two-person bench seats, and I was whispering to my seat-mate. The teacher really chewed me out." Figure 32 shows Pauline among the other Amherst students that year.

Clarence started to school at the Amherst school in the fall of 1920, also at age five. He also apparently had a few problems at school. At one point Albina remarked that "the only thing that Clarence had learned in his first year at school was to chew on his shirt collar."

32. Amherst School. Front row: Pauline is second from left. (1918)

Despite their apparently less-than-enthusiastic approach to grade school, Pauline and Clarence each made it through the eighth grade at the Amherst school, after which they each attended Luray High School. When they went to high school in Luray, the family car-pooled with other parents to drive them there, as shown in figure 33. Pauline graduated from high school in 1930 and Clarence in 1932.

Although Roy and Albina each had completed school only through the eighth grade, they valued education, and made sure that Pauline and Clarence would not only finish high school, but would also go on to college as well. This was during the Great Depression when many were unable to afford college, but Roy and Albina saw to it that both Pauline and Clarence attended Kansas State College in Manhattan, Kansas in the 1930s. It was unusual for female students to attend any college other than a teacher's college at that time, but that did not deter Roy and Albina from sending Pauline to Kansas State. Albina and Roy drove Pauline and Clarence on most of their infrequent trips to and from Manhattan, although Pauline remembered taking the train there at least once.

33. High school car pool. Gladys Bratton, Eulah Bratton, Freda Cochrun, Pauline Crawford; driver Russ Cochrun. (Fall 1928)

The Morrill Act, signed by President Lincoln in 1862, allowed for the creation of land-grant colleges, and in 1863 Bluemont Central College in Manhattan, Kansas became Kansas State Agricultural College, the first operational school created under the Morrill Act. Renamed Kansas State College of Agriculture and Applied Science in 1931, in 1959 it became Kansas State University.

In 1887, the Hatch Act authorized the establishment of agricultural experiment stations. These experiment stations were scientific research centers to study problems of food and agriculture and related industries. They affiliated with the land-grant college of agriculture in each state. The Kansas State Agricultural Experiment Station began projects at Kansas State Agricultural College in 1887. Some college faculty members also worked at the Experiment Station and carried out research there. Both Pauline and Clarence majored in areas related to agriculture and rural life, and professors involved with the experiment station may have taught some of their courses.

Pauline graduated from Kansas State in 1934 with a degree in home economics. She was a member of Alpha Xi Delta sorority and the Women's Athletic Association there. After graduation, Pauline took a job in Hoxie, Kansas, and in 1938, she moved to a job as the first Home Demonstration Agent in Stafford County, Kansas. Home Demonstration Agents were county representatives of the Department of Agriculture, employed to teach different homemaking skills throughout the community. Pauline settled in the town of St. John in Stafford County, where she became heavily involved with the local 4-H youth programs and other community educational programs as part of her Home Demonstration Agent responsibilities.

Clarence graduated from Kansas State in 1937 with a degree in Agricultural Engineering, having been active in the Engineering Association, the American Society of Agricultural Engineers, and in Steel Ring, an engineering honor society. He was a member of

Alpha Tau Omega fraternity, and was active in the Reserve Officers' Training Corps (ROTC) where he was in the Scabbard and Blade ROTC honorary society. Two years of participation in ROTC was mandatory for male students at Kansas State from 1931 to 1965. However, Clarence chose to participate in the program for a full four years, and so was able to receive a commission as a Second Lieutenant in the Army Reserves. After Clarence graduated from college, he was sure he wanted to be a farmer, and he moved back home with his parents to help with the farming.

Farm Mechanization Leaps Ahead

Roy and Albina strove to reduce the costs of producing their wheat, as did most other farmers. The 1920s saw a rapid change in the amounts and types of mechanization the Crawfords applied to the reduction of these production costs. In the early part of the 1920s, most of their farming operations depended on teams of horses. Figures 34-41 show Crawford farming activities in different years, and indicate that the 1920s were a time of transition for them, from farming primarily with horse-drawn mechanized farm implements to more sophisticated mechanized farm implements pulled by gasoline-powered tractors. This gradual transition enabled them to farm more acres with less hired labor, and reduced the need to devote some of the farm output as feed for the teams of horses.

34. Horses and wagon in front of the Roy Crawford house. (Early 1920s)

35. Roy with horses and wagon. (1920s)

36. Pauline, Clarence, and Roy picking up the newly mown grass in the Crawfords' yard. (ca. 1920)

In the early 1920s, Roy was still harvesting his wheat with a binder, as the Crawfords had done since late in the nineteenth century. The binder cut and bound the stalks of grain into sheaves or bundles tied with twine, and dropped these bundles in the field. Other workers picked up the bundles and stacked them into shocks, placing several bundles together so that the bundles held each other up with the ears of grain off the ground to dry out. Then they laid a couple of bundles across the top, arranged to shed rainfall from the shock. Figure 37 shows some of the shocks made from the Crawfords' wheat harvest in 1922.

113

37. Clarence and Pauline with cut wheat stacked in shocks in the field and waiting for threshing. (1922)

In the late nineteenth and early twentieth centuries, use of the binder was standard practice on the prairie. However, binding, shocking, stacking, and threshing still required considerable hand and horse labor. Many plains farmers turned to the header ($200), which cut a wider swath, worked well in short straw, and required less handling of the grain. The header did not tie bundles but merely clipped off the ends of the stalks holding the heads of grain and conveyed those loose heads to a wagon moving alongside. This eliminated the binding and shocking steps, and so required less labor. The horses powering the header pushed it from behind, since if they were pulling the header, they would trample the wheat before the header cut it. Figure 38 shows that Roy had switched to using a header for his harvest by 1924. The header required a team of four horses to push it (figure 38), and the wagons required two-horse teams to pull each of them, so a significant amount of manpower and horsepower was still essential to handle all these operations.

38. A header cutting the wheat and delivering it to the wagon alongside. Horses and manpower performed all the work. (1924)

By the turn of the century, Harmon had begun hiring custom threshing crews to thresh his grain. At that time, steam engines requiring constant feeding with coal and water supplied the power for the threshing machine. By the 1920s, the threshing machines used on Roy's farm still utilized a similar technology for threshing, but now a tractor with a gasoline or kerosene internal combustion engine supplied the power. Figure 39 shows the custom threshing crew at work on the Crawford wheat harvest in 1925.

"Dad" Montgomery was the leader of the threshing crew Roy usually hired, and he and his crew had their own ritual. The first thing Mr. Montgomery would do when he showed up at the Crawfords was to look for Pauline. He always had a large juicy chaw of tobacco in his mouth, and when he found Pauline, he would plant a big wet kiss on her cheek, making sure to leave tobacco stains there just to embarrass her. His hearty laughter would drown out her ensuing shrieks—then he was ready to get down to business.

39. Threshing time with tractor power. (1925)

He had a cook shack so the women of the household did not have to prepare meals for his crew, and because of that, they grudgingly tolerated his behavior. Soon the rest of his crew would show up with the tractor slowly pulling the giant threshing machine to the place Roy had chosen. They also brought additional horse-drawn wagons to supplement the wagons Roy had. The crew needed several wagons to keep a steady flow of the wheat heads from the header to the threshing machine and of the threshed grain to the grain bins or to the elevator.

To minimize the chance of starting a fire, the crew parked the tractor powering the threshing machine well away from the stack of straw, like the situation with the earlier steam engine powered threshers. A lengthy belt transferred the power from the pulley on

116

the engine through perhaps 80 feet to another pulley on the threshing machine.

The internal operations of the threshing machine remained much the same as they were in Harmon's time, and various pulley, gear, and sprocket linkages drove all the moving parts. Keeping the threshing machine steadily fed with the output from the header and ensuring that all these processes proceeded smoothly and efficiently still required considerable skilled and semi-skilled manpower.

In addition to the threshing crew, Roy had several regular hired helpers to haul wagonloads of the threshed wheat to the granary or to the elevator, unload the wheat from the wagons, care for the horses, run errands, and in general deal with the myriad of tasks and issues that occurred during threshing. Roy helped with these tasks as well as overseeing the whole operation, but he was allergic to the wheat dust so during threshing time he kept a moistened sponge tied over his nose.

In the early twentieth century tractors began to replace horsepower for performing agricultural tasks such as pulling farm implements for plowing, planting, cultivating, fertilizing, harvesting crops, and hauling materials. Tractors based on steam boilers were heavy and lacked maneuverability, making them impractical for most farm tasks, so they were clearly not candidates to replace the horses and mules.

However, the commercialization of the internal combustion engine led to a reasonable alternative, and the first commercial gasoline-engine-powered tractors appeared in 1902. The farm tractors that evolved utilized powerful internal combustion engines to drive the wheels to provide forward motion. They could also transmit some of this power to the attached implement through a power take-off shaft or belt pulley.

The tractor in figure 39 is an Aultman & Taylor model 30-60 ($2,700) using its belt pulley to power the threshing machine. This tractor, made sometime between about 1915 and 1924, was a popular powerful model based on a gasoline engine.

40. Roy with a team of horses, and with tractor in the background. (Mid to late 1920s)

At first, many companies produced gasoline- or kerosene-engine tractors of widely varying designs, but by the early 1920s Ford and International Harvester began to dominate the market.

Roy's tractor seen in the background in figure 40 is a McCormick Deering 15-30 tractor ($1,300), vintage 1924-1926, made by International Harvester. This tractor burned kerosene rather than gasoline, since kerosene was less expensive than was gasoline in those days. The model 15-30 was a large and powerful model with steel wheels designed with steel lugs to provide traction, and may have been the first tractor owned by any of the Crawfords. However, it had not yet replaced all of Roy's horses.

Roy acquired a Nichols & Shepard pull-type combine (about $2,000) in the late 1920s (figure 41). This was mostly crewed by Crawfords and Hampls. These combines married the header and a threshing machine to form an integrated unit to carry out both operations concurrently. A small internal combustion engine on the combine powered all the internal motions of the header and the thresher parts. A tractor pulled the unit through the field, cutting a swath of standing wheat or other crop and filling a bin or bags with the separated grain.

During the 1920s, pull-type combines such as this became popular in Kansas; a little over 8,000 tractor-pulled combines were operating in Kansas in 1926, but by 1930, this number had risen to 27,000. These combines eliminated the labor-intensive step of transferring the cut heads of grain from the binder or header to the threshing machine, and eliminated the need for the many men, horses, and wagons previously required for this step. Prior to the combine, the typical threshing crew consisted of 15 or more workers, while a combine crew consisted of only six or seven workers to operate the tractor and combine and to haul the grain to the grain bins or to the elevator.

41. Nichols & Shepard tractor-pulled combine and crew. Tom Hampl, Pauline Crawford, Clarence Crawford, one of the King boys, Charlie Crawford atop the combine, Bill Hampl, and Roy Crawford. (1929)

By the end of the 1920s, Roy had already acquired most of the farming implements he would use throughout the 1930s, including a mower, rake, grain drill, plow, harrow, and combine. Since Roy now had a tractor that could pull them, he began adapting those implements for use with the tractor, thereby gradually phasing out the use of his teams of horses. However, for sentimental reasons he did keep one team until well into the 1940s, occasionally harnessing them to mow the yard or perform other simple tasks.

One additional implement that Roy acquired in the late 1920s or early 1930s was a one-way disc plow or disc tiller. In 1916, Henry Krause built his first one-way disc plow, called a one-way or a wheatland plow, in a small farm shop in Western Kansas. The one-way left more stubble and other residue on the soil surface to fight wind erosion, leading some to call it the "first conservation tillage tool." The Krause one-way cut the soil, stalks, and weeds loose and tossed the lot a few inches to one side, severing roots and preventing regrowth. Instead of completely turning over the soil, as the moldboard plow did, the one-way only turned the soil part way over, mulching and mixing the weeds and stubble with the topsoil to slow erosion.

Krause's one-way plow was extremely popular with Kansas farmers, allowing them to till their ground much more rapidly than they could with the moldboard plow and to leave the soil better able to retain the moisture needed for planting the next crop. Other manufacturers soon recognized the importance of this implement and began producing similar tillage tools of their own. The one-way Roy purchased was a Minneapolis-Moline eight-foot unit. Sales of one-ways were brisk throughout the 1920s and 1930s, but they really exploded in the late 1940s, reaching 10,000 units in 1946 and 15,000 units in 1947.

42. Farmer tilling a field with a one-way.

As more people settled the Great Plains and more farms came into production, farm products such as wheat became more abundant commodities in the overall economy of the country. This relative abundance led to the supply exceeding the demand and hence to a lowering of the prices obtained for these farm commodities. This meant that the farmers had to produce more of those commodities to be able to afford to buy the same products they been buying all along, and this increased production tended to drive the commodity prices down even further, leading to a vicious spiral. Farmers have had to battle these market forces ever since.

The first years of the twentieth century were generally prosperous for American farmers. In the early 1900s better weather, better farming techniques, and more extensive mechanization drove heightened prosperity. However, during the years of 1914 through 1918 the combined effects of World War I and the Russian Revolution drastically diminished food production across Europe. Governments conscripted millions of European farmers into the various armies, causing European imports of food from the Americas to increase dramatically. To compound the problem, 1916-1918 were particularly bad years for wheat production in Kansas.

Both effects contributed to significant increases in crop prices. The high prices convinced many farmers to plant more wheat, and the number of acres in Kansas planted to winter wheat rose from about seven million in 1910 to about 11 million in 1919. The desire to plant more acres in turn led many farmers to go into debt to buy more land and more modern machinery with which to farm it, causing a boom in the prices of farmland and machinery.

European farm production returned to normal quickly after the end of the war, and this, coupled with the huge increase that had occurred in American farm production led to large crop surpluses starting in the early 1920s. From 1920 to 1930, the total acreage planted to wheat in Kansas increased by 30 percent, resulting in about a 30 percent increase in wheat production. Over the same period, the population of the US experienced only a 16 percent increase, not enough to utilize all the extra wheat. Overproduction led to rapidly falling prices—the price of wheat dropped back below a dollar per bushel by 1921, putting a severe squeeze on those farmers who had gone heavily into debt to buy land or machinery at the previous inflated prices. Crop surpluses remained high and crop prices continued to stay low through the rest of the 1920s.

The direct effect of the war and its aftermath on the Crawfords and Hampls was not so dramatic. Since they had resisted the urge to go into debt to expand their operations, the Crawfords and Hampls did not enter the 1920s heavily burdened by debt. However, they were not immune to the effects of the rise and fall of the prices they could get for their crops and livestock.

By the 1920s, wheat was the main cash crop for most of the farmers in western Kansas, and this was true for the Crawfords as well. The income a farmer received from a wheat crop depended primarily on four factors:

- The number of acres of wheat the farmer was able to harvest
- The yield of the wheat crop per acre
- The price the farmer received for that wheat
- The cost of producing the wheat, including labor, machinery, horses, feed for the horses, and other miscellaneous production costs.

The price of wheat depended strongly on wheat production elsewhere in the US and the rest of the world. Wheat had become a commodity and individual farmers had virtually no control over its price. Wheat prices varied throughout the 1920s, but were lower at the end of the decade than at the start.

Crop yields depended on many factors such as the weather and whether there were pest or disease infestations, leaving the farmer with only limited control. The average yields for Kansas wheat fluctuated wildly during this decade, but there was no general trend of increasing yields. Thus, the farmer's main option for maintaining or increasing his income during a period of decreasing wheat prices was to increase the amount of wheat he planted, usually only possible by converting grassland to cropland or by buying or renting more land. The rapid advances in mechanization had made it possible for farmers to farm the additional land with little or no additional help, making this option attractive.

Roy and Albina chose to purchase more land, and expanded their holdings over time. In 1920, Roy bought the Harmon Crawford homestead quarter on section 34 from his mother Candus. In 1928, he acquired the north half of the southwest quarter of section 34 (originally part of the O'Connor homestead) from Ora Beatty, widow of Zina E. Beatty who had recently died, and the northwest quarter of section 34 (originally the Thomas Beatty homestead) from Russell M. Cochrun. At that point, Roy owned 480 acres in section 35 and 400 acres in section 34, a large increase from the 160 acres Harmon and Candus had claimed with their homestead. These holdings represented a significant expansion of the Crawford farm operations, and they included the original Crawford homestead quarter, keeping that historical piece of land in the family.

Great Depression and Dirty Thirties

The stock market crash in 1929 and the ensuing Depression left far fewer people able to afford bread, and the price of wheat dropped to $0.33 per bushel in 1931 and 1932. Ideal growing weather in 1931 led to a bumper wheat crop, and that, coupled with the drop in demand, created a huge oversupply that helped to drive the price down and keep it down through 1932.

The First National Bank of Luray continued to serve Luray until the spring of 1933, when President Franklin D. Roosevelt closed all banks during a historic "banking holiday." After this closing, the vice-president of that bank, John O'Leary, Sr., and a few others raised the money to open the Peoples State Bank in Luray in 1934. Roy Crawford was one of the initial contributors.

In the aftermath of the banking holiday, Luray was too small to warrant a national bank, so the First National Bank permanently closed, selling its remaining assets to the Peoples State Bank. Mr. O'Leary served as president and board chair of the new bank. The Waldo bank also closed during the Depression, and Waldo never acquired another local bank. The citizens of Waldo and the surrounding community had to go to Luray or elsewhere for their banking needs.

Despite the loss of the local bank in Waldo, both Waldo and Luray remained busy commercial centers throughout the 1930s, continuing to serve their respective farming communities. The population of Luray reached a peak of 475 in 1920, and then remained nearly constant throughout the 1920s. Through the 1930s, Luray's population dropped slightly, but the number of students graduating from Luray High School increased from 168 for the decade of the 1920s to 196 during the decade of the 1930s, another indicator of the continuing vitality of the community. The Waldo town population grew slightly throughout the 1920s, reaching a peak of 279 in 1930 and remaining near that level throughout the 1930s.

By 1933, the price of wheat was back up, but drought arrived and the dust storms started. The first severe 1930s dust storm to occur in Kansas was in early May of 1934, and the worst of these "black blizzards" occurred across the entire Great Plains on Black Sunday in April of 1935. These storms, coupled with the by now serious drought, severely reduced wheat yields through 1935. The reduced production during those two dustbowl years removed most of the surplus wheat from the market, and the price of wheat began to rise slowly in 1936, but the drought continued to keep yields low into 1940. Normal rainfall and consequently better wheat crops finally returned in the early 1940s.

The Dust Bowl days in the 1930s seriously affected Russell County Kansas, along with much of the rest of the Great Plains area. At times, the dust was so thick in the air that it was impossible to see more than a few feet ahead. During one of those storms, Roy lost his way going from the house to the barn. He eventually walked into the fence around the garden, and from there he reoriented himself and made it back to the house safely.

Delmar Hampl provided recollections about the Dust Bowl.

> *"Dust storms were an unforgettable experience for anyone who experienced them. It affected the Crawfords the same as anyone else—trying to stop the fields from blowing by furrow tillage, etc. Dust was so fine that it went thru shingle roofs, windows, doors. It drifted like snow everywhere. It was a health hazard as everyone had cow lots, pig lots, chicken lots, etc. and it was all in the air. I missed a year and a half in the first and second grades in 1935 and 1936 due to a kidney infection. The 'thirties' was a double-whammy for health problems— dust pneumonia and other problems from the dust, and poor nutrition from the Depression. Times were tough.*

We were all poor but didn't know it because everyone was in the same boat."

Sarah (Griffin) Hampl, Alex Jr.'s wife, wrote in her memoir:

"We would get up in A.M. after a night of high wind from South and dust all over the yard would be red. We had drifts of dirt just like snow drifts in winter. It would shift from day to day. I'll never forget the first dust storm. We had gone to Luray to attend a dance. Got as far as [the] restaurant and the dirt and wind was terrible, so bad we never tried to go the one block to where the dance hall was. About 2 in A.M. [the] wind went down and people went home. When we opened the door to go in the house the drift of dirt in front of the door was thicker than my high heels. All over the house was dust. Our cat just walked around and kept shaking his feet. As soon as we cleaned a place on top of the sewing machine, he jumped up there and wouldn't leave. This happened time after time. Farmers, us included, lost chickens, animals from dust pneumonia. In fact, a lot of people died of it, especially small babies who had problems."

One of the lesser-known plagues brought on by the hot weather and drought of the 1930s was the proliferation of black-tailed jackrabbits, which were eating the meagre crops that had survived, including the roots. To combat this, many locales held jackrabbit roundups. At the beginning of a roundup, people lined up around the four sides of a square, spaced 20 to 30 feet apart. They then gradually closed in, making noise as they walked and forcing the rabbits into a fenced area in the center. There they clubbed the rabbits to death. Clubbing the rabbits was preferred to shooting them, since this method eliminated the possibility of the people accidently shooting one another. Besides, the bullets would have been too expensive.

43. Jackrabbit roundup. (1935)

Despite the drought, dust storms, and Depression, Roy was still intent on expanding the farm. On March 2, 1935 he bought the west half of the northwest quarter, the southwest quarter, and the west half of the southeast quarter of Section 32 (W½ NW¼ 32-11S-13W, SW¼ 32-11S-13W, and W½ SE¼ 32-11S-13W) from the heirs of Uriah Terry for $4,000. This nominal 320 acres was mostly pasture land, and was about three miles west of the Roy Crawford farmstead.

The years of 1939 and 1940 were still years of extreme drought. During this period, there was too little rainfall to replenish the water in the Crawford's cistern, and the cistern, which supplied all the drinking water and wash water for their household, ran dry. Roy had to haul water to their house from their well, several hundred yards distant from their house, to supply enough for drinking and other household needs.

44. Roy with the dog Bob and a ten-gallon "cream can," off to the well for water when the cistern ran dry. (August, 1940)

After suffering through a dusty scorching hot Kansas summer, the cool mountains of Colorado could provide a welcome respite. Once the intense activity of harvest was over, Roy and Albina along with family and friends would often drive to the mountains for a few cooler days of camping and fishing (figure 45).

45. Camping in Colorado. (1939)

Agnes Hampl died June 25, 1935. After her death, Alex Sr. went to live with several of their children. He died January 3, 1940 while living with Albina and Roy.

46. Pauline, Roy, Albina, Clarence. (1938)

As the 1930s neared their end, Roy and Albina were still running the Crawford farming operations, which had survived the

131

Depression, drought, and Dust Bowl relatively unscathed and had even expanded. Pauline had been out of college for several years and had begun her new job as Home Demonstration Agent in St. John; and Clarence, having recently completed college, was helping his parents with the farming.

"New Deal" Government Farm Programs

The continued loss in the volume and price of exports beginning in 1920 had led to a decade of relative decline in the farmers' income and purchasing power even before the beginning of the Depression. The national economic crisis in the early 1930s struck hardest at the farm sector of the economy, and it was the worst crisis in the history of agriculture in rural America. By 1932 farmers were making less than one-third of what they had been in 1929, farm prices had fallen more than 50 percent, and foreclosures were the order of the day.

Many people in the United States were in dire economic and social straits by 1933. The Agricultural Adjustment Act (AAA) of 1933, part of President Franklin Roosevelt's New Deal, aimed to raise farm prices through artificial scarcity. To help accomplish this, the AAA paid land owners subsidies for leaving some of their land idle. The AAA also introduced the concept of "parity;" an index based on farm prices relative to the general economy during the period 1910-1914, a relatively prosperous time for farmers. Parity provided a measure of the current exchange value or purchasing power of farm products. A goal of Roosevelt's farm programs was to return farm commodity prices to levels at or near parity, which would give farmers a fair "exchange value" for their products in relation to the general economy.

The AAA benefitted most farmers, with farm income rising 50 percent from 1932 to 1935, but in 1936, the Supreme Court ruled key provisions of the law unconstitutional. Congress passed a new version of the act in 1938 that remedied these problems, allowing agricultural support to continue and adding a provision for crop insurance. Farm support programs that evolved from these original New Deal programs are still central to the federal support for farmers; the US Department of Agriculture's Farm Service Agency administers these programs.

One facet of the farm programs was the issuance of non-recourse loans, in which the commodity itself is security on the loan. The Commodity Credit Corporation (CCC) issued these and they served to relieve the market glut at harvest time and to stimulate farm purchasing power. Specific percentages of parity set the loan rates, differing for each type of crop. The farmer could then keep the loan amount and surrender his crop to the government, or if prices went up sufficiently, he could pay off his loan and sell his crop on the open market. The non-recourse loan allowed the farmer immediate payment for his crop, alleviating some of the cash-flow problems incurred at harvest time, when farmers were often unable to sell their wheat due to insufficient capacity at the grain elevators.

However, the market price was frequently below the loan price, in which case the farmers usually let the government take their crop as the collateral, rather than repaying the loan, letting the loan rate effectively set the minimum price for the crop. Therefore, the government began to accumulate enormous stores of agricultural commodities. Partially to compensate for this, in 1935 the Federal Surplus Commodities Corporation began large-scale distribution of surplus farm products to needy families and to schools for lunch programs.

In 1938, the Federal Crop Insurance Program, a public-private partnership between the federal government and insurers, began providing subsidized crop insurance for various farming hazards. Insurers approved by the Department of Agriculture sold crop insurance policies. These insurers collected premiums, issued policies, and paid claims, while the federal government acted as a reinsurer, meaning it insured the insurance companies. A typical policy covered yield losses that resulted from natural causes like drought, excessive moisture, hail, wind, frost, insects, and disease. It also covered revenue losses that occurred when the harvest price differed from the projected price.

The farmer paid a portion of the premium for this insurance and the government subsidy paid the rest, so participation in this program was often a good deal for the farmer. The Federal Crop Insurance Program has evolved over the years, and many farmers still take advantage of this program, which accounts for a significant fraction of the total government expenditures on farm programs.

The AAA achieved some success in meeting its stated objectives. Output restrictions, non-recourse loans, and other price support programs raised farm incomes above what they would have been in the absence of any action. They also led to some cost to taxpayers and to higher prices for consumers at a time when many households could barely make ends meet. However, the AAA efficiently transferred funds to low-income farm families, and the cost was probably lower than the government would have had to face if the flight of farmers from the farms to the cities had continued at its rapid pace.

In 1935, Congress passed the Soil Conservation Act, creating the Soil Conservation Service (SCS), a locus for pulling together all the information on the best methods of farming safely within the capabilities of the land. The SCS promoted conservation practices such as strip-cropping, as well as contour cultivation of fields and the use of detention and diversion by water spreading structures such as terraces to conserve water and limit soil erosion. They also promoted the use of organic residues to increase organic content of the soil, and proposed keeping organic residues on the surface, as in the case of stubble-mulching, to prevent wind erosion. Removing critically erodible land from production and returning it to permanent vegetative cover was also a high priority.

Under this act, the government paid farmers to participate in such conservation programs, but only if they developed and agreed to follow an acceptable overall conservation plan for the land in question. This act became the cornerstone of federal conservation policy, and while technology has changed through the years, these essential elements still guide the soil conservation program. In the 1930s, the Crawfords adopted as many of these conservation practices as seemed reasonable at the time. However, they did not participate in any of the formal SCS programs until the mid-1940s.

47. The Crawford outhouse. Note that Albina is wearing a hat and heels with her apron, a level of formality uncommon today. (Early 1940s)

Another part of the New Deal was the Works Progress Administration, later renamed Work Projects Administration (WPA). The WPA provided paid jobs to millions of unemployed workers, including unemployed farm workers, during the Depression. Most of its programs aimed at building up the country's infrastructure, including parks, schools, and roads, but it also had programs in other areas such as the arts, drama, music, and literacy.

One little-known WPA program, intended to improve farm sanitation, provided help to build privies. The Crawfords took advantage of this program and had one of the WPA privies built to replace the existing outhouse on their farmstead. The WPA provided the standardized plans and the labor to build the privy, while the land owner provided the materials. The Crawford outhouse had a concrete floor, a concrete pot with a lid, and a separate screened ventilation shaft. Delmar Hampl referred to that privy as the "classiest and best built outhouse in Russell County."

The Changing Economics of Farming

By the end of the 1930s, the Great Depression, the Dust Bowl, and the New Deal programs introduced as a response to those events had initiated changes in Kansas farms and farming practices. The effects of those changes began to become apparent in the 1940s, and continued to play out through the remainder of the twentieth century and beyond.

Many of the smaller and/or under-capitalized farms became unsustainable by the mid-1930s, forcing the owners to sell their land, often to well-capitalized farmers. Most transactions were a bit more complicated than this, with banks foreclosing the small farms and selling them at auction or by other means, but the results were still the same. The net effect was that most of the land from those small farms ultimately ended up enlarging other existing farms. This exodus from the farms and the corresponding increase in the size of the remaining farms persisted at a rapid pace, although slowing somewhat toward the end of the 1900s.

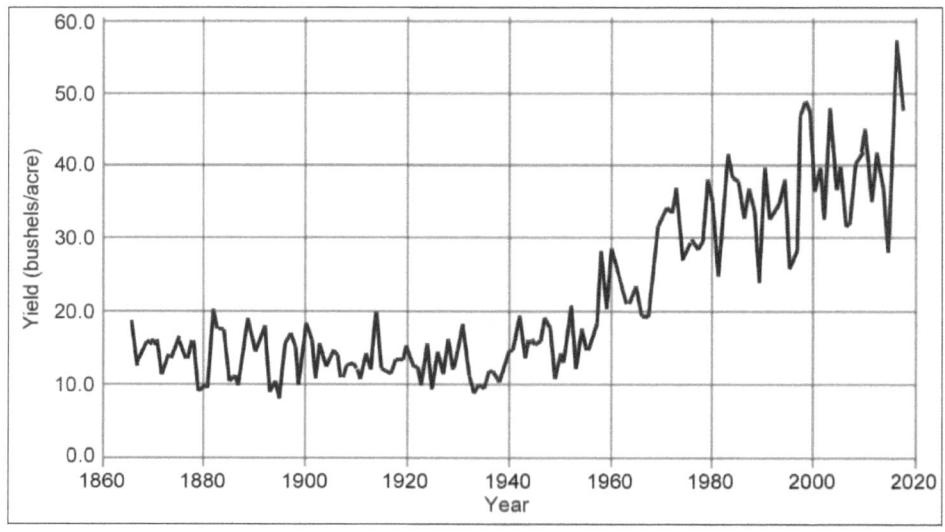

48. Kansas average wheat yields.

In the early 1940s, the average annual yield of the wheat crop began a steady rise that continued through the remainder of the twentieth century and through the early part of the twenty first century, as shown in figure 48. New varieties of wheat, mostly developed at the K-State and other agricultural research stations, led to roughly half of this increase in wheat yields. Improved management practices, in part based on the government conservation programs and other lessons learned from the Dust Bowl days, led to the other half of the observed gains.

These improved practices worked together with the new wheat varieties and included tillage choices, moisture conservation, type and timing of fertilizer application, and the decision whether to apply herbicide or insecticide. The agricultural research stations and the Farm Service Agency of the Department of Agriculture both provided guidance and assistance for farmers to improve those management practices.

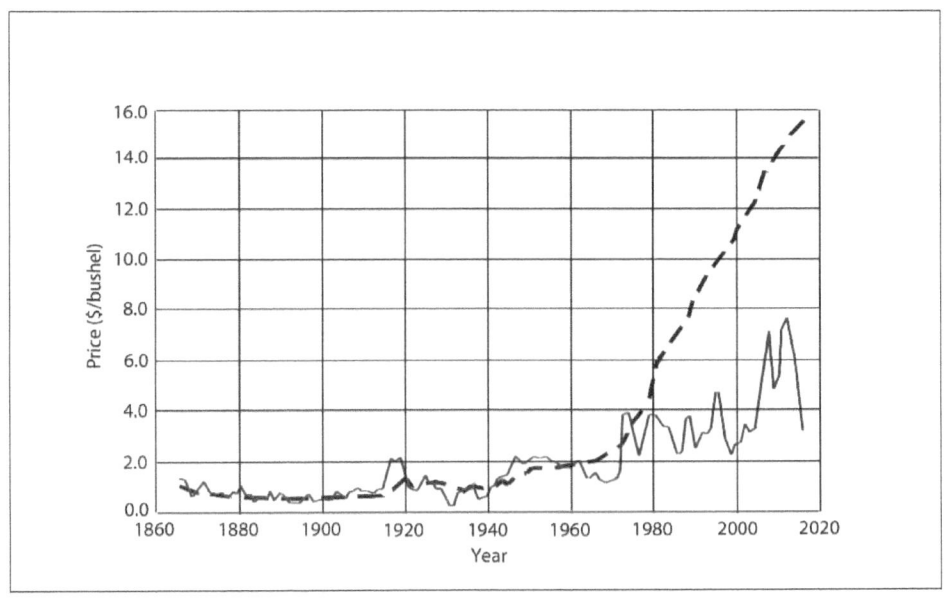

49. Average prices for Kansas wheat compared with Consumer Price Index: solid line—wheat price; dashed line—Consumer Price Index normalized to price in 1900.

Unfortunately, there has not been a corresponding increase in the real market price for wheat. As shown in figure 49, the average price of wheat in Kansas fluctuated around a dollar a bushel from 1866 to about 1940, except for a short period where it jumped to about two dollars per bushel due to World War I. By 1940, the price started to rise, averaging around six dollars a bushel by the mid-2010s. However, this gain was largely illusory, because as figure 49 shows, the cost of living rose at an even higher rate over this period. Even with the combination of yield gains and price gains, the income from a wheat crop could not quite keep up with the cost of living.

A change in government economic policy based on knowledge from the Depression led to most of this rise in the cost of living. The government began to take a more active role in management of the economy, and began to manage it to try to eliminate the periods of deflation, which tended to be periods of economic hardship for the general population. The net effect was that by about 1940 there were no significant periods of deflation and only periods of varying magnitudes of inflation, causing the Consumer Price Index (CPI) to rise steadily. That remains the situation to this day, with the government currently aiming to achieve an average annual rate of inflation of about two percent.

Not long after the start of the twentieth century, Roy and Albina became the owners and operators of the Crawford farm. Their tenure spanned a period from about 1908 to the late 1940s, during which they expanded the farm from the 480 acres that Harmon and Candus had ultimately owned to the 1440 acres Roy and Albina would leave to their children during the 1950s and 1960s. The continued rapid technological advances in farm mechanization, including the transition from horsepower to tractor power, made it possible for Roy to farm many more acres productively with only minimal help. Good farming practices and avoidance of debt enabled Roy, Albina, and their family to survive and even prosper during the Great Depression and the Dust Bowl in the 1930s.

Roy and Albina had done their part, maintaining and expanding the Crawford farming operations through troubled times, and then leaving those operations and that farm in excellent shape for the next generation of Crawfords.

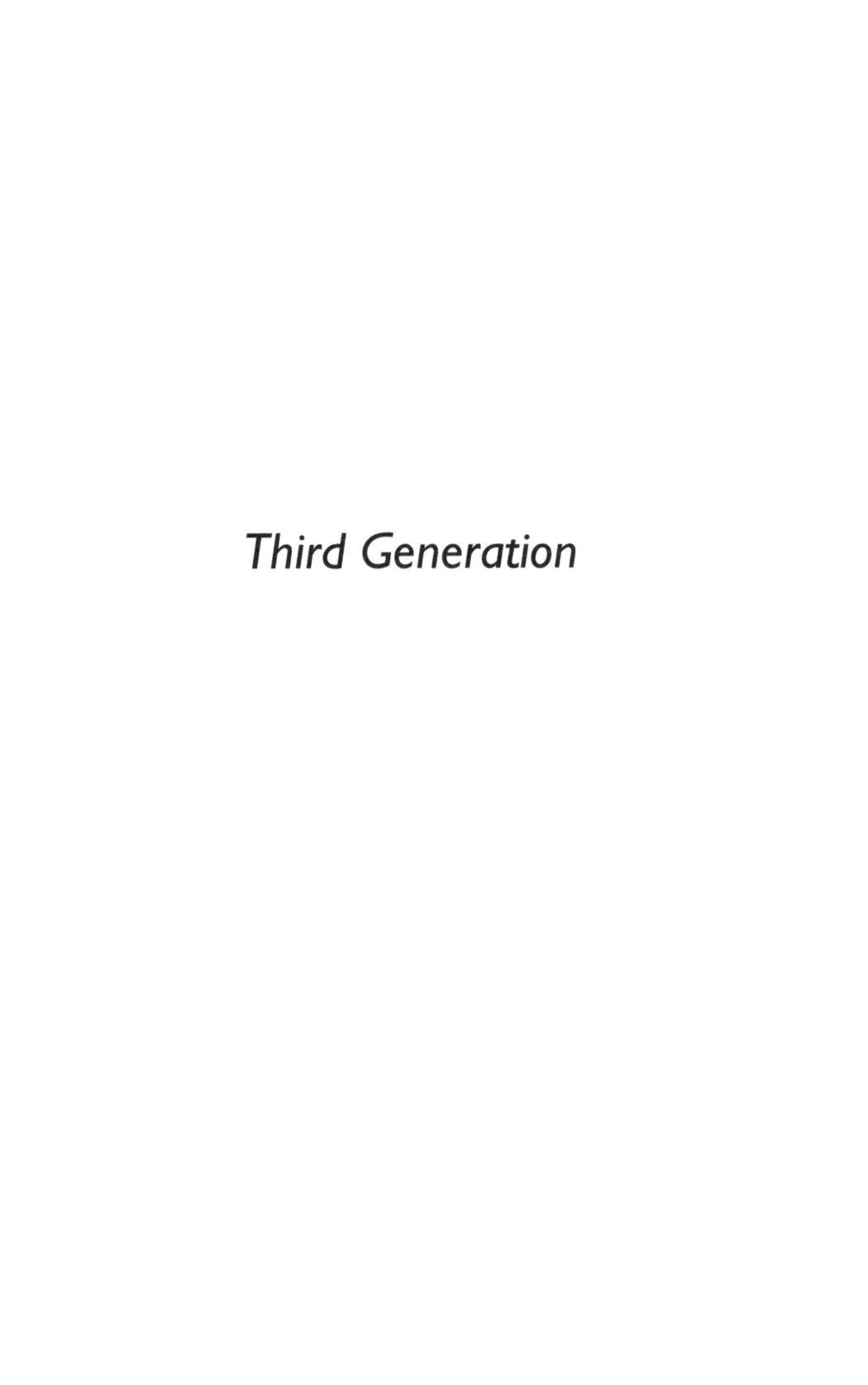

Third Generation

Preserving and Improving the Land

War Intrudes

Nine-year-old John Chegwidden looked back wistfully as his ship slowly left the Truro harbor, realizing that it might be months or even years before he again set eyes on this familiar piece of land. John, my great grandfather on my mother's side, was born in 1849 in Baldhu, a village near Truro in Cornwall, England. When he was nine, his uncle, who was the captain of a British merchant ship, took him as an apprentice to study seamanship as an alternative to a typical Cornwall future as a tenant farmer or a miner. John remained with his uncle for three years until he had finished his apprenticeship, and then at age 12, he enlisted for a term of five years in the British navy as a gunman.

When his term in the British navy expired, he signed for another term of five years at sea, but this time with the United States Navy as an experienced head gunman. At the end of that term, he was 22 years old and 13 of those years had been spent at sea. Tired of the sea, he decided to head to America where he expected to find other opportunities. His service in the US Navy enabled him to become a citizen of the United States almost immediately, and he sailed to the US on the Inman Line from Liverpool in 1871.

50. John and Mary (Lee) Chegwidden. (1886)

He soon went to work in the mines in Ishpeming, Michigan, an ironic but perhaps not surprising choice of jobs. In Michigan, he met Mary Lee. Mary was born Mary Trana in September 1851 in Stenkjaer, Norway. The Lee family adopted Mary and she immigrated with them to the US in 1874. John and Mary, my Chegwidden great-grandparents, were married in Negaunee, Michigan, in February 1875.

In 1876, shortly after the birth of their first child William, John Chegwidden and family traveled from Michigan to Kansas, where they homesteaded in Russell County roughly nine miles southeast of Dorrance, Kansas. Henry Lee, Mary's brother, traveled to Kansas with them, and homesteaded less than a mile from the Chegwidden farm. John later purchased that land from Henry and built a stone house there, locally known as The House of Seven Gables, and that became the home for John and his family.

The military also played a role in Caspar Sechtem's arrival in America, but his experiences were very different from those of John Chegwidden. Caspar was born in December 1853 in Sechtem, near Bonn, in Rhineland Prussia. When he was about 18, he ran away from home to avoid the compulsory military service in Prussia, and eventually stowed away on a ship bound for America.

This was at the beginning of the Franco-Prussian war, so it was not surprising that Caspar went to these lengths to avoid the draft. What was surprising was that soon after Caspar arrived in the United States in 1872, he enlisted in the US Army, serving in Company G of the Fourth US Cavalry stationed in Oklahoma Territory from 1872 to 1877. Most likely someone discovered him as a stowaway, and the US government gave him a "choice" between serving in the US Army or deportation back to Prussia.

After serving his five-year term in the cavalry, Caspar traveled north from Oklahoma Territory until he found work as a farm hand in southeast Russell County, Kansas. He later homesteaded there. In March 1882, Caspar married Mary Essig who was living nearby with her brother Conrad Essig. Mary had immigrated in 1881 from Weissach, near Stuttgart in Prussia. Caspar and Mary, my Sechtem great-grandparents, established their home and raised their family on the land Caspar had homesteaded.

51. Caspar and Mary (Essig) Sechtem. (ca. 1890)

The Caspar Sechtem farm was only about five miles from the John Chegwidden farm, and Sophia Sechtem, one of the daughters of Caspar and Mary, married William Chegwidden in April 1906 in Dorrance, Kansas. At first, Sophia and William, my Chegwidden grandparents, lived near William's parents and farmed some of his father's land, but in 1910 William purchased a 320-acre farm south of Lucas, Kansas, and established that farm as the home of his family. William and Sophia had five children: Helen, Lucille, Harold, Gladys, and Arliss. Gladys would eventually become my mother.

52. William Chegwidden and Sophia (Sechtem) Chegwidden.
(1906)

The Chegwidden children all attended Pleasant Valley one-room rural school, and later went to high school in Lucas, about ten miles southeast of Luray. Gladys was one of the younger children from a poor farming family, but she had a quick mind and rapidly advanced through the elementary school curriculum. Because she was from the country and was one of the youngest in the class, she was timid and defensive when she entered high school. However, there she discovered debate, a field in which she could excel, and gained enough self-confidence to work her way through high school and enter college at age 16.

In the late 1920s and early 1930s, Gladys and Helen attended college at Fort Hays Teachers College, taking on babysitting and other jobs to pay their way. After graduating and then teaching for a few years, Helen married Bill Hampl (Albina's younger brother) in 1934. Bill had been renting some farmland from Roy Crawford, and was living in the frame house Candus had built on the Crawford homestead quarter. Bill and Helen settled into that house after their marriage.

Gladys completed her Bachelor of Science in Education degree in 1932, earning a lifetime teaching certificate. After college, she taught three years at a country school near Holyrood, Kansas for $75 a month. In 1935, she began teaching English, French, and Latin in Luray High School for $100 a month. As part of that job, she also coached all the plays, oversaw the forensics program, chaperoned the basketball girls, and was a class sponsor.

53. Clarence Crawford and Gladys (Chegwidden)
Crawford. (1939)

Helen and Bill introduced Gladys to Clarence Crawford during one of his trips home from college. Clarence and Gladys saw more of each other once he had finished college in 1937, and that eventually led to their marriage at the Methodist parsonage in Luray, Kansas in August 1939. Dad wanted to live on the farm and to farm some of Grandpa Crawford's land, so when he and Mother were to be married Helen and Bill moved from the Crawford farm to Gorham, Kansas. There Bill began operating a Standard Oil gasoline tank truck, delivering gasoline to the local farms and to local oil well operations.

Although Dad wanted to live on the farm, Mother's years in college and her subsequent years of teaching school gave her the self-assurance and the dignity she missed as a child. She had no desire to go back to living on a farm, which for her had been a painful experience. For her to be willing to marry Dad and for them to live on the farm, she extracted a promise that the farmhouse they lived in would have electricity, indoor plumbing including a bathroom with hot and cold running water, and all the other modern conveniences that she had recently been experiencing while living in Luray.

After their marriage, Mother had to give up teaching—the prevailing policy at that time was that a married woman did not need a job because her husband should support her, so the job should go to someone else who needed the paycheck to support themselves and/or their family. Because of this policy, she only taught at Luray High School for four years.

After Mother and Dad were married, they moved into the frame house that Candus had built and Dad began renting and farming some of his father's land. They had a 1926 Plymouth coupe named Gracie for transportation, a cat named Tommy to keep them company, and just enough furniture to set up housekeeping.

With no electricity and no indoor plumbing, that house was far from having the modern conveniences Dad promised to Mother. A cistern collected rainwater from the roof of the house by a system of eave troughs and downspouts, and a hand-operated bucket pump supplied household water from that cistern. Nearby were an outhouse and a dark and damp detached cave that served as a cellar/tornado shelter. Occasionally that cave would partially fill with water during particularly wet periods. A coal-burning stove supplied heat for the house.

Mother and Dad immediately began to modernize that farmhouse to include the desired conveniences. They did some wallpapering and painting, began laying linoleum on the floors, and initiated the effort to install an indoor bathroom.

War broke out in Europe in 1939. Initially the American public adamantly opposed our country's involvement in this war, but by early 1941, the US began preparing to enter this conflict if necessary. While in college, Dad had participated in the ROTC program, receiving his commission as a second lieutenant in the Army Coast Artillery Corps in 1936, but the peacetime Army had little need for more officers so he had remained in the Army Reserves. By the fall of 1940 America's entry into the war began to appear inevitable, and in the spring of 1941, Dad received orders to report for active duty. His active status with a new rank of first lieutenant began in April 1941, so he had to put his farming activities and his projects to modernize the farmhouse on hold for the duration of the war.

After Dad reported for his Army duty, Grandpa Crawford had to handle all the Crawford farming, making use of whatever help he could find. Mechanized farming had reduced the work load for farmers, but tilling the fields, planting the wheat, and harvesting the wheat on the nearly 700 acres of cropland Grandpa owned still required considerable effort. Hiring someone to help with the farming was difficult because many of the young men in the area had already departed for or soon would depart for military service. Grandpa had to rely on a few good workers who were still too young or were otherwise ineligible for the draft.

Despite the difficulty in obtaining farm help, Grandpa continued to add to the Crawford land holdings, purchasing 160 acres in section 32 (W½ NE¼ and E½ NW¼ 32-11S-13W) from John A. O'Leary in July 1944, and 80 acres in section 27 (N½ SW¼ 27-11S-13W) from the heirs of Leander Beatty, one of Thomas Beatty's sons, when it became available in June 1945.

After a month stationed at Camp Stewart near Savannah, Georgia, Dad and Mother moved on to his permanent station at Camp Davis, which was in a swamp north of Wilmington, North Carolina. They arrived there in early June 1941, and I was born in October 1941 while they were still living in Wilmington. Not long thereafter, the Japanese attacked Pearl Harbor on December 7, 1941. Within the next four days, the US declared war on Japan, Germany, and Italy, and circumstances had suddenly fully embroiled the US in this conflict.

Shortly after the attack on Pearl Harbor, Dad's unit moved to Langley Field at Newport News, Virginia, and a month later moved on to Fort Dix, New Jersey, to prepare for shipping overseas. Grandpa and Grandma Crawford made the trip to Virginia to help Mother and me return to Kansas at that time. In February 1942, Dad's unit embarked from Boston, headed for Australia on the Queen Mary liner along with many other troop units—8,398 troops and 905 crew aboard. After "40 days and 40 nights" at sea, the Queen Mary docked in Sydney, Australia, in late March 1942, having made a circuitous 19,000-mile trip to minimize the danger from German submarines in the Atlantic.

During his stay in the Southwest Pacific, Dad's battery of the 94th Coast Artillery Regiment spent time defending airfields near Cairns on the far northeast coast of Australia, facing the Coral Sea. They also participated in the landings on Kiriwina Island, and later at Saidor, on the north shore of New Guinea, in each case setting up and operating antiaircraft defenses around airfields constructed as the Allies advanced.

While on Kiriwina, Dad received a promotion to captain and assumed command of his battery, which by then had become Battery B of the 209th Coast Artillery Antiaircraft Battalion. By late spring of 1944, the air base at Saidor had largely served its purpose. The Allied advance soon progressed further west, and in September 1944 Dad received orders for rotation back to the United States. He arrived at Fort Leavenworth, Kansas, in early October, 1944. All told, he had spent fourteen months stationed in Australia, six months on Kiriwina, nine and a half months in New Guinea, and nearly two months at sea while transiting to and from Australia.

While Dad was overseas, Mother and I settled into a two-story gray frame house, the Pangburn house, on Kansas Street in Luray. From there, we could walk to a grocery store, the post office, a bank, the doctor's office, and other stores to obtain most of what we needed. In the early 1940s Luray was a bustling small town with a population of nearly 400 inhabitants surrounded by a sizeable rural area whose inhabitants also considered Luray to be their home town — the primary place where they shopped and did business.

Shortly after Dad arrived back in Kansas, the Army reassigned him to Fort Bliss, Texas, where he participated in the training of new antiaircraft units. Mother and I joined him in traveling to Texas, and we spent the next year living in El Paso. In September 1945, Dad received the order to begin processing for separation from the Army, and he completed this separation and departed from Fort Bliss on September 30, 1945. This was the last of his army activities, although his official discharge from service did not occur until early January 1946.

We packed and left for Kansas as soon as possible after Dad's tour at Fort Bliss ended. After we got back to Kansas, Dad rented all the Crawford land in sections 27 and 34 from Grandpa Crawford, in a crop-sharing arrangement. This came to a nominal 480 acres; roughly 460 acres of this were cropland and about 20 acres were pastureland.

Dad began farming on this cropland while the three of us settled into the house in Luray where Mother and I had lived during the war—just in time for my fourth birthday. Soon after we settled in Luray, Dad's sister Pauline Crawford married Dale Bookstore in December 1945. Dale was a farmer and rancher from St. John, Kansas. His first wife had died, leaving him with two young children: Thelma Dalene and Robert Dale. Pauline and Dale and the children settled on Dale's farm just outside St. John.

Our longer-range plans were to move to the farm, but the house there still had no electricity and no indoor plumbing and needed quite a bit of remodeling before it reached the standards Dad had promised to Mother. We spent most of the next year and a half living in Luray while Dad remodeled that house and got back into the operation of the farm.

Luray had remained prosperous throughout and after the war, supplying nearly all the goods and services required by the community. Most of the businesses were conveniently located on Main Street, and were only a block or two walk from our house. I remember a hotel that provided accommodations for travelers, and a restaurant and a cafe offered a choice to those wishing to dine out, although we rarely ate at either place. There was a Chrysler/Plymouth dealership that also sold International Harvester products and could supply a tractor or truck and a variety of farm implements. That dealership also had trained mechanics to service the vehicles they sold. A service station provided gasoline as well as some basic automobile maintenance services.

I occasionally got to go with my folks when they shopped at Joe Libal's grocery store. Opie Mowrey's drug store stocked a variety of personal supplies, and offered a soda fountain where I could often get Grandpa to buy me an ice cream cone or a popsicle. My folks banked at the Peoples State Bank, which provided banking services for the community, and we frequently walked to the post office to pick up or send mail. A telephone office staffed by a switchboard operator was only a short walk away.

The railroad provided daily passenger and freight service to Luray; the railroad depot was at the south end of Main Street. I don't remember ever riding on that train but I did like to watch the steam locomotive pull the train into and out of the station. Farmers could sell their grain or purchase many of their farming supplies at two different grain elevators located next to the railroad.

Dad bought most of the materials and tools needed for his remodeling project at the nearby Mack-Welling Lumber Co. hardware store and lumberyard, but since I was only four and apt to get into things I shouldn't, he usually did not take me with him when he shopped there.

The community held many functions in the American Legion Hall or in the Auditorium/Gymnasium building also located on Main Street. The latter building operated as part of the Luray school system, but the grade school and high school buildings were located about four blocks from the business district. Community members could also choose between two churches in town, the Methodist Church, and the Evangelical United Brethren Church.

My sister, Candace Ruth Crawford, was born in April 1946 while we were living in Luray. Living in town, close to businesses, was especially convenient for our parents with the new baby to care for.

Life on the Farm

Finally, in early 1947, the time came for us to move to the house on the farm. Dad had completed the massive remodeling project that included removing several interior and exterior doorways to improve the usability of some of the rooms, turning some of the space into an indoor bathroom, and building an enclosed back porch with a laundry room. The house now had three large bedrooms upstairs, and the downstairs contained the kitchen, bathroom, a small bedroom, a family room/dining room, and a large unfinished room that initially served as a junk room.

The new back porch had a stairway leading down to the storm cellar. Dad had refurbished the storm cellar, giving it concrete walls and floor and a concrete arched ceiling. The floor of the storm cellar was several feet below ground level. However, the interior of the cellar was tall enough inside for an adult to stand up since the roof of the cellar was above the level of the ground outside. From the outside, it looked like a little hill in our yard. The inside of the cellar was small, and since it contained our electrical generator and our water pump with its pressure tank there was barely room for the four of us to crowd inside. Fortunately, we never experienced a tornado, and when conditions looked bad, we just went to the back porch where we could be near the door to the cellar.

Dad had wired the house for electricity; a gasoline-powered generator mounted on a concrete stand in the storm cellar supplied the power. A plumbing system distributed water from the cistern to the kitchen, the bathroom, and a laundry sink located on the back porch. An electric pump and pressurized storage tank system, also located on a concrete stand in the storm cellar, provided the water pressure, and a water heater in the bathroom supplied hot water. The house now contained a kitchen sink, a modern bathroom with a flush toilet, tub, and sink, and a laundry space on the back porch; all but the toilet having both hot and cold running water.

Dad also had a cesspool constructed to handle the effluent from all the drains. A floor furnace provided the heat for the house, and it burned propane gas supplied from an external storage tank located in the shelter belt trees behind the house. The kitchen sported a propane-burning gas range for cooking and an electric refrigerator to keep food from spoiling, while the back porch contained a deep freezer for storing frozen meat and other foods. Few other farmhouses in the area had all these modern amenities at that time. Life here would bear little resemblance to the rudimentary living conditions experienced by Harmon and Candus when they settled on this same farm site in the late nineteenth century.

54. Our house on the farm after Dad finished remodeling it.

Our farmhouse was about seven miles from Luray, so there would be no more walking to the grocery store or the bank or playing daily with friends there—our family now had to drive seven miles over gravel roads to get to town. At least we had gravel all the way. Many rural houses in the county were reachable only by travel over dirt roads—and those dirt roads were nearly impassable for automobiles when wet. Since it was no longer practical to go to the grocery store every day or two, the deep freezer on the back porch became essential.

There was more than an acre of yard surrounding our house. The original shallow well dug by Harmon Crawford was about 200 feet from the house, and a large ramshackle barn and a chicken house were farther away. The well still didn't provide much water, but most of the time it was adequate to supply the nearby stock tank.

161

Before long, Dad tore down most of the old barn and hired a carpenter and his helpers to build a machine shed with concrete walls and a galvanized iron roof in its place. Dad used some of the stones that came from the original stone house built by Harmon in the concrete walls of the machine shed. He designed this machine shed to hold some of the farming machinery he intended to purchase and to provide him with a shop space and a multi-purpose concrete-floored space where he could store grain if necessary.

Dad suspended a swing inside the shed and later put up a basketball backboard, hoop, and net in the area with the concrete floor. I played many games of H.O.R.S.E. with friends there. Candace and I occasionally used this multi-purpose space for roller skating as well.

Shortly after we moved to the farm, we acquired a Jersey milk cow. Dad had converted a small building behind the machine shed into a shelter for milking the cow and for storing the feed for her. We also acquired chickens and housed them in the chicken house; gathering the eggs soon became my daily chore. A garage built on the site of the original stone house using some of the stones from the original stone house protected our family car from the elements.

Our telephone was a wooden box hung on the wall, like the one shown in figure 55. This box, still based on early 1900s technology, had a plastic mouthpiece to speak into and a receiver that the user held up to the ear. The mouthpiece contained a carbon microphone that turned the spoken sounds into electrical currents from a battery located inside the wooden box, and each receiver on the network utilized an electromagnet and a diaphragm to convert these electrical currents back into sounds.

55. Wooden box wall telephone.

All the people in the area were on the same party line and could listen in on anyone's phone calls. A crank on the side of the telephone operated a magneto to generate a current, causing a bell to ring on each of the phones in the system. Each house had its own code—our code was two long rings and three short ones—and everyone was supposed to answer only the calls that were for them. Of course, some people listened to all the calls just for entertainment, so the sounds of someone else breathing heavily usually accompanied any use of the phone and discouraged casual phone usage. Calls to someone not on that party line required a call to the operator who could make a connection to other systems— our operator was in Waldo.

Thunderstorms were common and most of the limited rainfall we got came in downpours. Copper wire, strung on telephone poles along the roads, linked the phones in the neighborhood. This extensive network of copper wire made an effective connection to the electrical activity associated with the thunderstorms, so when there was lightning, arcs would sometimes flash out a foot or more from the telephone. I wouldn't walk anywhere close to the phone at such times; this was occasionally a problem since I had to walk down a narrow corridor past the phone to get to the bathroom.

We were on a "Rural Route," so our mailing address was RR2, Luray, Kansas. Laverne Oswald was our mail carrier for most of the time we lived on the farm. He delivered mail to our mailbox next to the road, and picked up any outgoing mail we might have had.

Although we had indoor plumbing, we used water sparingly because the only water supply for the house was the cistern, and that only received the rainfall collected from the roof. We were not supposed to leave a faucet running or to use more than about an inch of water in the bathtub. Showers were unheard of. Our well barely supplied enough water for the animals. Deep wells provided plentiful potable water in some areas, including in Luray and Waldo, but in most of the Amherst area, the deep wells only brought up salt water that was unfit to drink.

The Harry King family lived not quite a half a mile to the north of us, and Grandpa and Grandma Crawford lived about a half mile to the northeast of us. The Zenas Beatty family and Lloyd Beatty and his wife Bertha Maude lived about a half mile to the south of us. Lloyd was Zenas and Ethyl Beatty's son; Lloyd Beatty and Bertha Maude (Reese) Beatty had been classmates of Dad in high school, and Lloyd had also been in grade school at Amherst with Dad. Bertha Maude and Lloyd were wonderful neighbors, generous, kind, and fun-loving, and we visited with them frequently. These were our nearest neighbors, and none of these neighbors had any children near the ages of Candace and me.

As far as I was concerned, this isolation from other children was the worst part of the move to the country. However, in the summer of 1947, the Ivan King family moved into a house on the Harry King farm. Ivan was one of Harry King's sons and had been one of Dad's classmates at Amherst and Luray. He and his wife Margaret had two children: Newton (Newt), who was a year older than I; and Kay, who was a couple years younger than Newt and so was at least nearer to Candace's age. Finally, there were some children nearby to play with.

In 1936, Congress passed the Rural Electrification Act, a part of FDR's New Deal, creating the Rural Electrification Administration (REA) and directing it to provide low-cost loans to companies and community cooperatives to build the infrastructure to provide centrally generated electricity to rural homes. The war delayed this process, so the infrastructure (power lines) from the Smoky Hill Electric Cooperative, formed in 1945, did not reach the area around our farm until the late 1940s.

As did nearly all the local farmers, Dad and Mother signed up for the REA service. A tall power pole with a transformer and a yard light on it was soon standing next to our garage. Overhead electric wires from that pole provided electricity to the house and to the machine shed. Since Dad had already wired our house for electricity, it was a simple matter to tap into the REA power rather than the generator. The availability of the REA power made our lives a little more convenient, making it possible to operate additional electric appliances at the same time, to operate electric tools in the shop in the machine shed, and to use that shop after dark. The yard light was also a welcome addition, since it opened new possibilities for use of the yard in the evening. The REA electrification made a much greater impact on most of the other farmsteads in the area, including that of Grandpa and Grandma, since many of them had previously been unable to operate electrical appliances or in some cases even electric lights.

Dad had a love for the land that had begun in his childhood, and he was intent on applying the best soil conservation practices on the land that he farmed. One of his favorite statements that summarized his philosophy was "This land has been very good to our family and we need to be good to it." He repeated that statement often. Shortly after we moved to the farm, he began pushing Grandpa to participate in the Soil Conservation Service conservation programs, and on April 8, 1948, Dad as the operator and Grandpa as the owner agreed to a soil conservation plan covering Grandpa's land in sections 27, 32, and 34. They designed this plan in conjunction with the United States Department of Agriculture Soil Conservation Service.

This plan required the construction of broad-base terraces draining into grassed waterways on part of the cropland. The Soil Conservation District aided in the design and layout of the terrace system, but the farmer was responsible for the construction and planting of the waterways and the construction of the terraces. Dad first hired someone with earth-moving equipment to shape the waterways to provide good drainage and to eliminate any mud holes. He then planted the waterways with a suitable grass mixture to protect them from significant erosion.

After Dad had established the grass in the waterways, a surveyor laid out the terraces according to the land contours, maintaining the desired drainage gradient. Dad then used the 8-foot Moline one-way to move the dirt to build and shape some of these terraces, and hired a person with a road grader to build and shape others.

Each of these terraces had a channel on the uphill side, and the soil removed from this channel made a raised berm on the downhill side of the channel. The terraces followed the contour of the land with a slight downward gradient along the channels and into the waterway. This design forced the water to flow slowly along this gradient, giving it more opportunity to soak into the soil, rather than running rapidly down the hillside and eroding some of the soil as it would in the absence of the terraces. Once completed, those terraces effectively eliminated nearly all the hillside erosion in those fields, and the moisture they retained added to the productivity of the land. The terraces would wear down after a few years of farming, so Dad occasionally had to use the one-way to restore their shapes.

The conservation plan also called for construction of a large pond in the pasture in section 32. The completed pond provided plenty of water to allow that pasture to support more cattle, making it more valuable as a rental property. Grandpa had that pond stocked with catfish, giving him and Grandma a nearby place to fish—I and various other members of our extended family frequently went there to spend a few hours fishing as well.

Dad closely followed the progress of the crop development research at the Kansas State Agricultural Research Center at Manhattan, as well as the research at the Agricultural Research Center—Hays, a branch of the Kansas State Agricultural Experiment Station. The Hays experiment station, established early in the twentieth century, focused on development of wheat varieties adapted to western Kansas, while the main research center at Manhattan concentrated on varieties adapted to central and eastern Kansas. The Kansas Agricultural Experiment Stations developed many of the wheat varieties commonly planted in Kansas. Sometimes Dad obtained seed wheat from one of the experiment stations, which he would carefully plant to avoid mixing with any of his other wheat. He was also careful when he harvested this wheat, since he wanted to be able to use some of it and perhaps also sell some of it as seed wheat for the next year's crop.

56. The Luray Grade School building (constructed 1908) on the left and High School building (constructed ca. 1910-1912) on the right. The old Luray water tower is in the background. (Early 1950s)

The nearby Amherst rural school had closed in 1946, so Candace and I attended grade school in Luray. The Luray school system did not offer a kindergarten, so my first encounter with school was when I entered the first grade in September of 1947. Someone had to drive me seven miles there every morning and pick me up from there every afternoon. Fortunately, we were able to car pool with the King family, each set of parents driving on alternate weeks. Whenever the roads were muddy from a rain or

169

when there was a heavy snowfall, the driver had to put chains on the car tires to provide enough traction to avoid being stuck in the mud or snow.

The Luray grade school was a big step up from the one-room schoolhouses Dad and Mother had attended—the Luray Grade School building, built in 1908, was a two-story four-room limestone schoolhouse with only two grades per room! Each classroom had one teacher to teach both grades sharing that room. A basement, apparently an afterthought to the original building, contained the rudimentary but functional bathrooms.

Each student sat at a desk with a smooth wooden writing surface (smooth unless some previous student had carved initials into it), a groove to keep pencils and pens from rolling off, an ink well for an ink bottle (we still used fountain pens with ink—messy), a shelf for books and supplies, and usually an ink stain somewhere. Underneath the desktop and underneath the shelf were great places to stick used chewing gum wads—we found a lot of evidence that earlier students had already discovered this.

Slate blackboards dominated the front wall of each classroom; many lessons required blackboard work with students writing there with chalk. Frequent erasing of the blackboards created chalk dust that lingered in the surrounding air for much of the day, but living in a Kansas farm community we were already accustomed to having dust in the air most of the time.

There was a break for recess in mid-morning and another in mid-afternoon and an hour break for lunch. A cafeteria in the basement of the nearby high school building provided hot lunches.

The government subsidized these hot lunches by providing quantities of surplus farm commodities originating from the government's farm programs. Most of the students, including me, ate the hot lunches, which our parents paid for monthly, but a few students brought their lunches or else went home for lunch.

Weather permitting, we played outside during the recesses and during the remainder of the lunch hour after we finished eating. There was open space around the school, and there were swings, a seesaw, and a merry-go-round in one corner. The lower grades frequently played on this equipment, but sometimes we would play tag or a game we had made up. The upper grades usually had a "work-up" softball game going, where players advanced through the positions until they came up to bat. Once they made an out, they went into the outfield and all the other players moved up one position. Often there was an ill-defined activity involving a football occurring in another part of the playground at the same time.

A year or two after I started to school, Dad won election to the Luray Board of Education. He served on this board for six years, part of the time as board president. In addition to his stint on the school board, he was active in the Luray branch of the American Legion and in the Luray Lions Club. Mother taught an adult Sunday School class in the Luray Methodist Church.

At the start of the 1950s, almost no one had a TV set, and we had to look elsewhere for entertainment. We did have radio though, and we played many board games. Monopoly was popular, as was Pollyanna, a Parker Brothers game with rules similar to Parcheesi. Checkers and Chinese checkers were other popular games. There were also games like 20 Questions that didn't require any boards or other equipment. Newt King and I would have lengthy games of Monopoly, often ending only when one of us was losing and began accusing the other of cheating.

Newt and I also played a lot of ping-pong. We didn't have a ping-pong table, but we had an old pool table in our junk room. We managed to string a net across the pool table and use it as a ping-pong table at times. There were some interesting bounces when the ping-pong ball hit one of the side cushions on the table. I'm sure we made up some rule to cover those situations, and then argued about the rule until we forgot what it was.

When I went to Newt's house, he didn't have either a ping-pong table or a pool table. However, they did have a junk room in their house too, and it contained an old dining table that we extended with a leaf. We usually opened that table fully, put the leaf in to fill about half the opening, and played ping-pong on that. Of course, the gaps in the table surface led to the need for special rules and the ensuing arguments as well. As was our junk room, their junk room was unheated and we had to wear our coats in the winter. I learned to play left-handed as well as right-handed so I could keep warming one hand at a time in my coat pockets.

57. Gladys, Clarence, Candace, and Kent. (1949)

Drop-in visits with the neighbors remained an important form of entertainment. One of my favorites was when our parents would take us to visit Alex (Jr.) and Sarah Hampl. Alex and Sarah were among the first in the neighborhood to have a TV set. Candace and I could sit in front of the set trying to watch the shows while the adults visited. The marginal reception made it a

173

challenge to figure out just what was going on in some of the shows, but the sheer novelty of the TV kept me entertained. Card games were also family entertainment. Our family loved to play slapjack and spite & malice. Canasta was also big in the 1950s, but it required more than two players to make it interesting. Sometimes folks would start a canasta game at family gatherings.

We did a lot of reading. Luray had no library and the school had only a few books to loan to students, but we did have other reading material. When Amherst school closed, Dad bought their Book of Knowledge encyclopedia set at the auction. I read a variety of articles, poetry, and stories from that encyclopedia. In addition, the mail carrier delivered the *Luray Herald* and the *Salina Journal* newspapers, and our family subscribed to several magazines as well: *The Saturday Evening Post, Life, Colliers, Redbook, Readers Digest*, and a few more. Grandpa Crawford also subscribed to the *National Geographic* magazine, and stacks of back issues found their way to his attic for future generations to enjoy. These books, newspapers, and magazines were a source of news, literature, humor, and information—different from many magazines today.

Thunderstorms were a common occurrence in the spring and summer. I liked to sit out on the front porch with an unobstructed view of the clouds, where the rest of the family often joined me. Those clouds grew darker and darker as they formed and circulated. I could look for the lightning and then count slowly while waiting for the thunder to arrive. Five seconds between the lightning and the thunder meant the lightning was about a mile away. If the sky started to take on a greenish tinge it meant the

174

storm was about to get violent with hail and possibly a tornado; then it was time to leave the front porch and head to the back porch where the family could be close to the entrance to the storm cellar.

Often there would be entertainment in town. Of course, there were the concerts and ceremonies associated with the schools. Anticipated events included an annual Alumni Banquet for alumni of Luray High School (or for anyone else who wanted to buy a ticket), and the Luray United Methodist Church sponsored an annual fish fry as a money raiser. The Luray Lions Club occasionally sponsored or staged events to raise funds for some of their projects. A few times, the Merl Markley farm implement and Chrysler-Plymouth dealership hired a group to come in and put on a show, usually country and western music complete with corny comedy. The school auditorium, located in the Main Street business area, served as the venue for all these events.

Saturday nights were still a big social occasion when all the farm families put on their clean clothes and went into town to shop and to socialize. The adults would do the shopping and visit with friends while the kids could run freely up and down Main Street playing games, listening in on the old men standing around embellishing stories with one another, or just fooling around with groups of their friends.

Until I was about ten, one of the first things I would usually do on Saturday night was to find Grandpa Crawford, because he would take me into Opie Mowrey's drug store and buy me a cherry Coke or an ice cream soda. After having satisfied this highest priority item on my agenda, I might then spend a little time at

Opie's magazine rack, reading the new comic books until Opie chased me away. Then I would go find the other kids, and we would play tag or other games or make up other activities, usually constraining ourselves to the two-block long business area of main street.

My parents would often end up visiting with the Wineingers, who ran a service station and hotel, and I would go meet them there when it was time to go home. I often ran around with the Wineinger boys, Keith and Bruce, so this was convenient. Those were simpler times, and the largely-unsupervised freedoms we had then would be unheard of for most children today.

Occasionally we would go to town early and eat supper in Dave Tate's Café. I liked to eat there, and especially liked the large framed reproduction of a painting entitled "Custer's Last Fight" that dominated the café decor. Anheuser-Busch, the makers of Budweiser Beer, supplied this "painting" of the battle often referred to as Custer's Last Stand, and it had their logo in the corner. The focus of this painting was Custer dressed all in gold with his long yellow-gold hair flowing with the action. Custer's gold outfit positively glowed, and it grabbed my attention completely. When I looked a little more closely, I could see that he was heroically defeating the many Indians that surrounded him. After seeing that painting numerous times as a young child, it was a long time before I realized that rather than being the triumphant hero, Custer had mismanaged the battle and had been killed in his inglorious defeat—hence "Custer's Last Stand." One of life's little disappointments.

Wheat Harvest

World War II led to a sharp rise in commodity prices and a six-year period of strong farm income starting in 1939. The large stocks of wheat, cotton, and corn resulting from the Commodity Credit Corporation takeover of defaulted price support loans became a military reserve of crucial importance as war once again curtailed food production in Europe, and even more so after the United States entered World War II. Concern over the need to reduce the buildup of government stocks of commodities changed during the war and post-war period to concern about attainment of the production to meet war and post-war needs. The demands of war led to a steady increase in the acreage planted to wheat in Kansas, rising from 12.4 million acres for the 1940 crop to 14.1 million acres for the 1945 crop. The price of wheat also showed a steady increase during the war years. Yields during those years varied, but were beginning to climb from their 1930s levels.

The large number of deaths, degree of economic devastation, and displacement of populations in Europe and Asia during World War II retarded the postwar recovery of food supply systems in those areas. Therefore, agricultural prices remained relatively high, and US agriculture avoided the slump that had characterized most prior postwar periods. Between 1946 and 1949, the Kansas acreage planted to wheat continued to increase, from 14.1 million acres to 16.2 million acres. The Korean War beginning in 1950 gave another two-year boost to prices.

Wheat harvest was always a time of excitement and hard work. Much of the farm income depended on how good the harvest was. In the summer of 1948, Dad hired one of the crews of custom cutters that came through the area to harvest the wheat and haul the grain to the elevator in town. They had three self-propelled combines and a crew of about ten men, and spent several days cutting our wheat. The going rate for custom cutting in the late 1940s and early 1950s was typically $3.00-$5.00 per acre and $0.10-$0.20 per bushel. In that same time, wheat sold for about $2 per bushel and a good yield might be 20 bushels per acre, so the cost of having it custom cut was not negligible but was usually acceptable.

58. Custom cutters gather their trucks and combines at the Amherst schoolhouse. (1948)

Mother had to provide the noon meals for all these men as well as for the rest of the family. To help with this she hired three of her cousins, Lois, Myrtle, and Phyllis Cline. Beginning early in the morning, they prepared heaping platters of fried chicken, mashed

178

potatoes, and one or more vegetable dishes each day for the noon meal for the crew. The crew also devoured multiple apple, cherry, and lemon meringue pies per meal. Fortunately, we didn't have to provide sleeping accommodations for the crew, most of whom stayed in the hotel in Luray.

Shortly after the wheat harvest, the summer of 1948 brought its share of sorrow. My grandfather William Chegwidden died in August 1948, at the age of 70. He was buried in the Wilson cemetery. After William's death, Grandma Sophia Chegwidden moved to a house in Lucas. Sophia (Sechtem) Chegwidden died in August 1950—she was buried in Wilson next to William.

In the fall of 1948, Dad bought a new ton and a half Ford F-5 truck (about $1,500) and a new Massey-Harris Model 21-A Self-Propelled Combine (about $5,000). He had a special bed fabricated in Salina and put on the truck. This bed had a thick wood floor with removable wooden side panels used for hauling a load of grain and other loose items. It also included a blower that he used to transfer grain from the truck into a granary. Removing the side panels facilitated hauling other things that did not need tight containment.

The Massey-Harris Model 21-A was by far the most popular self-propelled combine model at the time. It cut a 12-foot-wide swath through the wheat field, and meant that Dad could cut his own wheat and not have to pay the custom cutters to do it.

However, he still occasionally hired custom cutters to cut some of his wheat when it was urgent to get it harvested.

59. Dad in his combine cutting the wheat in one of his fields. (ca. 1950)

Although Dad could now do the harvesting himself, he still needed help to do it efficiently. He needed one person to operate the combine and another to drive the truck. Whenever the combine bin got full, the operator emptied the grain into the truck using the auger and spout attached to the combine bin. In addition, when the truck got full the truck driver had to drive it to town to deliver the wheat to one of the elevators.

Drivers usually took the wheat to one of the elevators in Luray. If they were full, the driver might go to an elevator in Waldo instead. However, storing the wheat at home was sometimes necessary because the grain elevators occasionally ran out of storage space.

While the combine was cutting the wheat, other fields might need cultivation, and this would require an additional person to run the tractor. Grandpa and Dad together could adequately carry out most of these functions, but occasionally it was necessary to hire extra help.

In 1950, Grandpa gave all his land in sections 27 and 34 and 240 acres of his land in section 32 to Dad. These 720 acres contained approximately 470 acres of cropland and 250 acres of pasture. Dad was also renting and farming Grandpa's section 35 land, excluding the house and farmstead; roughly 380 acres of cropland and 90 acres of pasture. This freed Grandpa from most of his farm responsibilities, and allowed him and Grandma to indulge their love of travel by spending the next few winters in warmer climes.

In 1949 or 1950, they bought a trailer house, and they spent the winters in McAllen, Texas. Later, they began wintering in Punta Gorda, Florida. Several other family members and Amherst neighbors also wintered in those same areas, so they had a thoroughly enjoyable time.

Regulars on these treks included Harvey and Mary Bean and Chester and Nellie Beatty. Mary Bean was Grandma Albina Crawford's sister, and Harvey Bean was a son of Jacob E. Bean who was one of the early settlers in the Amherst area. Chester Beatty was a son of Samuel Beatty and was a grandson of the Thomas Beatty who homesteaded just west of the Harmon Crawford homestead. Because they still liked to be close to farm activities, Grandpa and Grandma returned to Kansas for the spring through fall months, but they occasionally went on shorter trips during those months whenever they felt the urge.

As Grandpa tapered off his participation in the farming, Dad needed to hire someone else to help with the farm work. I was still too young to drive the tractor, so Dad employed Dale Whearty, a young man whose home was near Manhattan, Kansas, and Dale spent the next few summers living with us and operating either our tractor or our combine, as needed.

Leonard Bean, one of Dad's cousins, had recently taken over the dealership for Case tractors in Luray, previously owned by his father. Dad bought a new Case model LA tractor ($1,650) from him in 1951. Now Dad was no longer dependent on using Grandpa's tractor. However, when Dad had Dale or someone else helping him his Case and Grandpa's International Harvester tractor might both be working, each on a different part of the land Dad was farming.

Helping with the Farming

In the late 1940s, Dad bought eight Hereford beef cows, with the intent of raising their calves to sell. Occasionally he kept one of these calves until it was full grown, and then had it butchered. This supplied more than enough meat to fill our deep freezer.

A few years later he bought a bull (named Larry Domino based on his pedigree) to go with the cows. These cattle lived in the pasture and corral just south of our house. That pasture contained a pond with water for the cattle, but that pond usually ran dry for at least some parts of the year. When that pond was dry, the cattle drank from a stock tank in the corral adjacent to the water well. I often had to pump water for the cattle from the well to the tank when the level in the tank got too low. The well had a hand pump, but using that pump to fill the stock tank was a tedious task that I did not enjoy. Somewhat later Dad replaced this pump with an electric one.

This small beef herd provided a nice extra source of income, but it brought with it additional responsibilities. The cattle could graze in the pasture during the spring through the fall, but Dad had to cut and store additional hay for the cattle during the summer and then put it out for the cattle to eat in the winter when there was nothing to graze.

To keep the cattle from leaving the pasture or corral we needed to maintain the fences. Sometimes even in the summer the grazing in the pasture was poor, and then we had to drive the cattle over to Grandpa's pasture where additional grazing was available, and that required maintenance of the fences there as well. In the winter, we had to put out hay or other food to supplement what the cattle could get from their meager grazing opportunities. Often when the weather got bad in the winter, we had to herd the cattle to Grandpa's barn for shelter. And always there was the need to make sure the cattle had a plentiful supply of water.

During the first years that we had beef cows, Dad cut hay and stored it in haystacks, much the way farmers had done it for decades. However, now a tractor and a truck replaced teams of horses. First, Dad cut the grass with a sickle mower, letting the hay fall right where it was cut so that it made an even covering on the ground. Then he pulled a rake behind the tractor, collecting the hay into small piles each about three feet high.

After that, he parked the truck next to one of these piles and we pitched the hay onto the truck with pitchforks — this is where I came into the process. This job usually required one or more persons on the ground pitching the hay onto the truck, and one person on the truck using a pitchfork to move the hay into secure positions so that it wouldn't fall off when the truck moved. Dad showed me how to handle the pitchfork and how to orient the hay, and then it became my job to position the hay on the truck. I was about eight or nine years old at that time. We repeated this process to pick up the hay from each of the piles.

When the truck was full, Dad drove the truckload of hay to the place where he wanted to build the haystack, and we used pitchforks to move and stack the hay onto the haystack. This time the person on the stack (me), had to be especially alert to place the hay in the right orientation so that it would hold the stack together as we added more weight on top. When the stack was complete, I rounded it on top and took care to orient the top hay so that it would force any rain to run off rather than getting the hay inside the stack wet. Dad checked periodically to make sure I was doing this correctly.

One drawback to this whole process was that the rake collected more-or-less everything in its path. Sometimes this included snakes, and having a pitchfork full of hay that included a snake, possibly a rattlesnake, could be startling. Fortunately, this was a rare occurrence, and I don't recall this happening more than once or twice during the time we were putting up hay this way. After a couple of years of haystacks, Dad started hiring one of the neighbors to cut and bale the hay. Picking up and stacking the bales was hard work, but it was a much simpler and more efficient process than building the haystacks from loose hay.

I learned to drive when I was 12, driving our truck in the pasture while Dad and Newt King loaded hay bales onto it. Although only a year older than I, Newt was a lot bigger and stronger than I was—strong enough to handle the bales. That left me to handle the driving. At 12, I could not legally drive on the public roads, but it was legal for me to drive around in our pasture. The truck had a stick shift, and it took me a while to get the hang of operating the clutch smoothly. I had to stop frequently to allow the loading of more bales, and every time I started for the next group of bales, the truck would give a mighty jerk as I engaged the clutch too fast.

Newt was stacking the bales on the truck, so he was riding on the back of the truck when I made one of my worst starts. Unprepared for my jolt, he lost his balance and fell off the truck. Fortunately, he was unhurt, and perhaps even more fortunately, my driving improved after that. By the end of the day, I had mastered the techniques of the clutch and brake, and my starts and stops were much smoother. However, it took much longer than the end of that day before I finally heard the last wisecracks about my driving.

By the spring of 1954, I was regularly milking the cow, and that summer I raised a steer and two pigs as 4-H projects. Those projects convinced me that I really didn't enjoy raising livestock, and I thought that I would rather be doing fieldwork instead. By the summer of 1955, Dale Whearty was no longer working for Dad, and I began operating the tractor pulling a one-way to till the fields. From that summer on, I was able to contribute significantly to our fieldwork, driving the tractor for many of the tilling and planting operations. Our chickens were all gone by that time, so Candace did not get the experience of gathering eggs, but she was helping Mother with the cooking, the laundry, and other household tasks.

Our tractor did not have an enclosed cab, so when I was operating the tractor, I was out in the open and had only my clothing plus a straw hat for protection from the elements. Driving the tractor to till our fields gave me a much greater appreciation of our farm, and a better understanding of how Dad and Grandpa could so love farming. There was something special about being out in the open, away from everyone else, having a full hemisphere of clear sky above and a 360 degree view uninterrupted by nearby trees or buildings. I felt like I was in some sense communing directly with nature.

It was easy to imagine the feelings Dad and Grandpa must have had, knowing that much of the land they surveyed from atop the tractor was theirs by virtue of their own efforts and those of their ancestors. Working in partnership with this land and with nature, they had created and left their mark on a farm that could support their families and provide for generations yet to come. I began to understand what Dad meant when he repeated, "This land has been good to our family, and we must be good to it!"

However—nature frequently intruded on such reveries, as each turn of the tractor wheels lifted a little more dust and grit into the air and the ever-present Kansas winds drove those particles into my eyes, nose, mouth, and through my clothing to cover my entire body. While I was communing with nature, nature was embracing me in return. By the end of the day, covered in dirt, sweat, and some grease from the tractor or implement, I felt that I was doing a man's work now.

All that dirt also meant that I would have to clean up in the washroom on the back porch before entering the house. Cleaning up in the washroom involved an experience of a different kind— Lava soap! Lava soap was an olive drab bar soap that contained pumice and felt like sandpaper. It didn't so much wash the dirt and grease off, it was more like sanding it off, but it sure was effective in removing almost anything from the skin, sometimes even the outer layer of the skin itself.

Dad was always looking for ways he could conserve and improve the farmland, and in the late 1950s and early 1960s, he bought several new tillage implements that left even more of the stubble on the surface than did the one-way, a desired conservation and soil improvement practice.

The implement receiving the most use was a blade plow (undercutter). The blade plow had several overlapping single-piece V-blades, each about 30 inches wide, mounted on widely spaced shanks. The shanks created only a few soil openings, and so limited moisture loss. The tractor pulled the undercutter with the V-blades traveling through the soil a few inches below the soil surface, breaking up the soil and cutting off the roots of any weeds that might be growing at the time. The undercutter left up to 75 to 95 percent of the stubble and other residue on top of the soil, much more than was the case with a one-way. This residue helped to conserve moisture as well as to protect against erosion, so the undercutter largely replaced the one-way as the tillage tool we used and that I pulled behind the tractor.

At the age of 15, I got my chance at the pinnacle of farm operations—operating the self-propelled combine. From then on, that became my favorite farm task. Operating the combine was a job that required all my attention. Not only did the combine need steering along the correct path around the field to capture all the wheat heads efficiently, I had to raise and lower the header of the combine depending on the height of the wheat stalks and the profile of the ground.

This job was extremely dirty and noisy, and potentially dangerous. Like our tractor, our combine also did not have a cab to protect the operator (me) from the sun and the dirt and noise; I sat out on a platform exposed to all of it, so goggles and a straw hat were absolute necessities. Wheat dust was everywhere, and it was far more irritating than ordinary dirt.

Once the header cut off the wheat heads and some of the stalks, a series of augers, conveyer belts, and other contraptions moved the heads and stalks through the rest of the combine, carrying out the necessary operations to separate the wheat grain from everything else. The machinery then moved the grain into the grain bin and spread the "everything else" out the back of the combine.

All this machinery involved numerous sprockets, chain drives, belts, and pulleys to transfer power to the various moving parts, and these utilized more than 30 bearings that I had to grease two or three times a day. Crawling under the combine was the only way to reach some of these bearings. A grease gun and a bucket of grease accompanied the combine to each field where I worked, and grease smudges and a thick layer of wheat dust covered me.

In addition to steering the combine and controlling the height of the cutting platform, I had to listen carefully to the sounds made by all those moving parts. Any abnormal sound usually meant that there was a problem to diagnose and deal with—often a blockage in the system. This required shutting down the combine, opening the covers to access the offending part, and then digging out by hand whatever weeds or straw had created the blockage. Operating the combine made me feel like I was a responsible adult doing an adult job, at least in this activity.

Farm, Town, and School

In the early days, manure spread on the fields augmented the supply of many of the basic nutrients required by the growing crops. When tractors replaced the horses, that change eliminated one of the major sources of manure for use as a fertilizer. One of the main nutrients required by the crops is nitrogen, and scientists began to realize that nitrogen applied as a chemical fertilizer could partially compensate for the missing manure, significantly increasing crop yields.

Scientists knew the processes for manufacturing ammonia and a few other nitrogen-containing compounds by the beginning of the twentieth century, and since nitrogen was the principal ingredient in most explosives, the advent of World War II triggered a rapid development and refinement of these processes. By the end of the war, factories were producing huge amounts of ammonia, and after the war was over, much of this production capacity switched to the manufacture of fertilizer, especially important, as the war had generated a pent-up need to restore food supplies in Europe and the United States.

Potassium and phosphorus are also important plant nutrients. The soil in the Amherst area required the addition of phosphorus to produce high yields of wheat. Research efforts from federal and state institutions led to the development of guidelines for the optimum types and amounts of fertilizer to apply and the best times to apply it. Plentiful deposits of potash from around the world provide a ready supply of potassium for fertilizer. However, phosphates, mined in Florida and the major source of phosphorus, are not a renewable resource. Dad frequently complained that it was very shortsighted to sell any of this US phosphate supply overseas.

The extensive use of fertilizers led to the development of more advanced fertilizer application equipment. Sprayers became an essential implement for modern farming, and evolved to sport longer and longer boom lengths as time went on. Farmers removed many fences and other obstructions from the fields to avoid damaging these long booms.

By 1960, farmers in the US were applying more than seven million tons of these chemical fertilizers per year. Dad began applying some chemical fertilizers by the late 1950s, and continued routine use of them as needed from then on.

By the early 1950s Europe and Asia were well on the way back to feeding themselves, and the Korean War ended in 1953. Both these developments led to a reduced demand for farm commodities, and coupled with the increased yields across Kansas and the rest of the country, created new surpluses. The general wheat surplus kept the price well below $2.00 per bushel throughout most of the 1960s. The government kept the non-recourse loan rate above the market price, giving farmers an incentive to produce more than markets would absorb, and the government-owned surplus grew rapidly. Some of the surplus farm commodities went to support the school lunch programs and other domestic food subsidy programs, but the surpluses piled up much faster than these programs could utilize them.

The Agricultural Trade Development and Assistance Act of 1954, better known as Public Law 480, established the Food for Peace program. The Food for Peace program provided a way to transfer the remaining surpluses outside the country, with humanitarian donations or sales of commodities at concessionary prices helping the poor in other countries.

The Agricultural Act of 1956 established the Soil Bank, similar in some respects to programs of the 1930s, to reduce the supply of agricultural products by taking farmland out of production. Part of the program was a conservation reserve in which the government paid farmers to designate certain cropland for the reserve and put it to conservation use. However, typically the farmers put only the poorest land into the soil bank, so although it helped preserve and improve the land, this Act did little to reduce the massive grain surpluses.

In 1951, Kansas experienced some of the worst flooding since the "Great Flood of 1844." The 1951 flood began when 11 inches of rain dumped down on Big Creek near Hays (25 miles west of Russell) in less than two hours. Big Creek flows into the Smoky Hill River, so this deluge caused substantial flooding not only in Hays but also in towns and countryside downstream along the Smoky Hill. Heavy rains in June and July extended this flooding across a broader area, resulting in the most severe flooding occurring in many of the areas in eastern Kansas and in Kansas City, Missouri. In all, floodwater covered about two million acres, damaging or destroying 45,000 homes.

Damaged cropland in the flood plains remained unusable for several years. Our farm was not in a river or creek floodplain, so did not suffer direct flood damage. However, the continued heavy rains caused the stalks of the ripe wheat to collapse, a condition known as lodging, so that harvest became an effort to manipulate the combine to retrieve the wheat heads lying close to the ground, a delicate and demanding operation that eventually salvaged enough of the wheat to achieve a partial but disappointing harvest. Fortunately, Dale Whearty was still operating the combine for us at that time, so that delicate task did not depend on my yet-to-be-developed combining skills.

The next few years after the flood of 1951 brought an extended period of drought to Kansas. This drought started in Texas in the late 1940s, but spread across the Great Plains in the early 1950s, resulting in severe drought across Kansas from 1952 through 1956. In terms of rainfall, runoff, and ground-water recharge this drought was more severe than the drought of the 1930s. Once again, there were dust storms depositing dust drifts in the fence rows in some areas. However, following the 1930s experience farmers had instituted improved soil conservation practices such as returning some of the poorest land to grassland, farming on the contour, the construction of terraces, and leaving stubble and other residue on top of the soil when tilling. These considerably reduced the number and severity of dust storms in the 1950s.

The 1950s drought resulted in a general and widespread reduction in ground water levels. In many places, the ground water levels fell below the bottoms of existing wells, resulting in the need to haul in water for livestock and domestic use. Dad acquired a water tank that he could transport on his truck, and used that to haul water that he purchased at Luray where there was a deep-well water supply. He had to haul water to replenish our cistern and stock tank several times during this drought period, and he occasionally hauled water to one or more of our neighbors as well.

As if a record flood followed by a record drought were not enough, the 1950s also brought a record blizzard and ice storm. The Blizzard of 1957 occurred March 23-25, starting with rain and sleet, which left a thick coating of ice on all the power lines and telephone lines. The subsequent two days of high winds and blowing snow broke power lines and telephone lines in multiple places, shattering many of the supporting poles in the process. It took over two weeks to restore electrical power and telephone service to our farm. During that period, we used kerosene lamps to light our house.

After the first couple of days, it became possible for Dad to drive to Luray. At that point, he took Candace and me to Luray where we stayed in the Wineinger hotel for two weeks until we had power on the farm. The schools in Luray had remained operational, so this enabled us to go back to school after minimal interruption. Meanwhile, Dad had loaded his gasoline-powered electric generator on his truck and spent time driving from neighbor to neighbor, supplying power at sufficiently frequent intervals to prevent the neighbors' freezers from warming up and spoiling their frozen food.

The Wineinger hotel in Luray was usually full during harvest time when the custom cutting crews were in town, but it generally wasn't crowded the rest of the year. Most of the ground floor of the two-story hotel building served as the living quarters for the Wineingers. One room on the bottom floor served as the lobby for the hotel while the upstairs held eight or ten small bedrooms surrounding a communal bathroom. The two Wineinger boys, Keith and Bruce, each had one of the hotel rooms as their own room, and I stayed in one of the other hotel rooms. I think Candace stayed downstairs with the two Wineinger girls, Patricia and Beverly. The Wineingers provided breakfast and dinner for Candace and me.

The 1950s drought resulted in significantly reduced crop yields during many of those years, although the 1950s still had a few years when average wheat yields reached new highs above 20 bushels per acre. In some of the drought years, the fall planting did not result in a good stand of wheat. Like many other farmers at the time, Dad started replacing some of the poor wheat with milo, a type of grain sorghum, planted in the spring to augment the farm income. He harvested the milo in the late fall, and this evolved into a more regular alternation between wheat, milo, and summer fallow, with different sequences for different portions of the cropland.

The population of Luray was nearly constant throughout the 1950s, starting with 350 people in 1950 and finishing the decade with about 330 in 1960. Eldon Hampl opened an electrical and plumbing sales and repair shop in part of the former Pospishil Opera House building. Eldon was another one of Dad's cousins, the son of Alex Hampl Jr. and Sarah (Griffin) Hampl. In the mid-1950s, Dave Tate retired from operating his restaurant, and for a few years Ivan and Margaret King took over that business. Opie Mowrey retired from operating the drug store, and Virgil and Ethel Siemers remodeled it and reopened it for business. Aside from those few changes, the businesses on Main Street remained much the same as they were in the 1940s.

In 1951, a roller-skating rink took over the upper floor of the old Pospishil Opera House on Main Street. For a little more than a year, this was one of the best places for young people to congregate in Luray. Unfortunately, the first floor directly below the roller rink was a vacant apartment. In 1953, the new music teacher and his family rented that apartment, and the noise from the roller rink overhead would have made living in this apartment intolerable. This ended the short but glorious era of the Luray roller rink.

In 1958, Governor George Docking appointed John A. O'Leary as Commissioner of Banking for Kansas, and John moved to Topeka to function in that position. John had to resign from his position as president of the Luray bank, leaving his son, John A. O'Leary, Jr. to take charge of the Peoples State Bank in Luray as president. John Sr. remained as State Banking Commissioner until 1968, when he resigned that position to assume the position of Director of the Federal Reserve Bank of Kansas City, Missouri. Meanwhile, John Jr. had become one of Luray's most active boosters, and Russell County named him Man of the Year in 1961.

Up until the mid-1950s, the grain elevators in Luray had been the iconic sheet-metal clad wooden structures ubiquitous across the plains. Those elevators had limited grain storage capacities, and in 1956, the Osborne County Co-op Association replaced one of those elevators with a much larger modern slip-form-poured silo-like concrete structure, and installed modern grain-handling technology in this new facility. The other elevator in town was unable to compete with this technologically advanced large-capacity facility and it soon went out of business.

The increased capacity of this new elevator also attracted business from elevators in the surrounding towns, especially from those in Waldo. The population trends in the two towns dramatically demonstrated this effect. The population of Luray dropped by only seven percent during the 1950s, while that of Waldo decreased by 21 percent. The construction company that was building these new grain elevators was also marketing composite slip-form-poured concrete and steel water towers. Each tower was a spheroidal steel tank supported by a slip-form poured hollow concrete column. Near the end of the decade, the town of Luray was sufficiently confident about sustaining its population that it paid to replace its aging steel water tower with one of this new type.

During the 1950s, Candace and I progressed through school in Luray. I graduated from high school in the spring of 1959. In the fall of 1959, I entered college at Kansas State University. Candace began the first grade at Luray in the fall of 1952, and entered Luray high school in the fall of 1960.

In Kansas, one could get a restricted driving permit at age 14, and for me this occurred in the fall of 1955, shortly after I had begun my first year of high school. Once I was able to get my driver's permit, I got a car. Then I could car-pool with Newt and Gail Hall, who was a year older than Newt and lived along our route to school. This eliminated any need for my parents to drive me to or from school from then on.

In addition to educating the children, the schools also served to connect the rural areas and the town into a unified community. That community understood the need for the schools, and tacitly assumed that the existence of the schools meant that the schools were educating their children. This left the community members free to concentrate on what was most important about the schools—football! The people of the Luray community closely followed the fortunes of the Luray High School Panthers football team, and a winning season could sustain community pride until the next season rolled around.

While I was at Luray High School, there were typically about 45 students in the freshman through senior classes. Roughly half of these were boys. Luray still played the regular style of football with eleven players per team, so nearly all the boys in the high school signed up for the football team, and all that signed up would automatically be part of the team and have a good chance of playing on the first team.

Luray was in class BB, the designation for high schools with fewer than 60 students, and many of the nearby class BB schools had already switched to a modified form of football with only six or eight players per team. Thus, Luray frequently had to look far afield to find enough class BB or class B (fewer than 150 students) 11-man teams to fill their nine-games-per-season schedule. Luray ended up playing some larger schools such as Kensington, Lebanon, Lenora, Logan, and Holyrood that were 50 to 100 miles away.

Luray usually was successful at football, but there was a period while I was in high school when we were remarkably successful, amassing a string of 22 consecutive wins in the years 1956-1958. Even though local farming was still suffering from a drought, the success of this team uplifted the spirits of the whole community, and the public regarded that football team as something special. The Peoples State Bank of Luray even treated the entire football team to a steak dinner at one of the restaurants in Russell. Unfortunately, archrival Lucas beat us in the final game of the 1958 season, payback for the game that capped off our perfect season the previous year. This rather inglorious ending to our 22-game winning streak left those previously uplifted community spirits somewhat diminished.

Once Dad completed his World War II service in 1945, he and Mother took over most of the responsibility for the Crawford farm operations. They modernized the frame house Candus had built on the homestead quarter, and we moved into it in the late 1940s, making it once again the center of the Crawford farming activities.

Dad's emphasis for the farm was on conserving and improving the land. To this end, he began the process of terracing all the Crawford cropland to minimize water erosion and to increase the moisture content of the soil, relying on guidance from the Farm Service Agency. He also adopted more conservation-friendly tillage implements such as an undercutter to minimize wind erosion and to provide an additional boost to the moisture content of the soil. By the end of the 1950s he was well along on his program of improvements on and preservation of the Crawford land.

Also, by the end of the 1950s, Dad and Mother had finished shepherding me through twelve years of school in Luray and were sending me off to college. The farming activities were under control, and life appeared to be continuing as normal in the Amherst and Luray communities they considered their home. They were blissfully unaware of the big changes that were on the horizon.

Fourth Generation

END OF AN ERA

Leaving the Farm

In the fall of 1958, I entered my senior year in high school faced with the prospect of a daily drive to and from school in Luray by myself. No more carpool—there were now no others in the area with the same needs to stay after school for sports practices. Gail Hall, an earlier carpool member, had graduated in 1957, and Newt King graduated from Luray High School in the spring of 1958.

Soon after Newt graduated the King family left the Luray area, leaving the King farmstead standing empty. Their departure also led to the closing of the restaurant they had been operating in Luray. Newt went on to college with a football scholarship at the University of Kansas in Lawrence. Ivan, Margaret, and Kay moved to Topeka, about 35 miles from Lawrence, and Ivan began operating a service station in Topeka. I felt Newt's absence strongly throughout my final year in high school, especially during my daily drives to and from school.

I graduated from Luray high school in 1959, one of a class of 17 seniors, the largest class in Luray High School at that time. I spent that summer at home helping with the farm work, and in the fall of 1959, I entered college at Kansas State University in Manhattan, where I majored in physics. I came back to help with the farm work during the summer of 1960, but that was the last time I would participate in the work on the farm in any significant way. In 1961, I had a summer internship in the Kansas State physics department, and in 1962, I had a summer internship in physics at the Savannah River Laboratory at Aiken, South Carolina.

While at college, I met Charlotte Anderson who was from Russell, and in January 1963, we were married in Russell. We both graduated from college in the spring of 1963, and spent the summer of 1963 in the Washington, D.C. area where I had a summer internship at the Johns Hopkins University Applied Physics Laboratory. In the fall of 1963, I started graduate school in physics at Princeton University in New Jersey, while Charlotte taught math in Bound Brook, New Jersey. It was clear that neither of us was on a path that would lead back to life on a farm in Kansas.

I know that Dad was disappointed that I would not be following his example and eventually taking over the farm, but to his credit, he let me decide the direction I would head and did not try to steer me one way or the other.

In 1958, Grandpa and Grandma Crawford celebrated their fiftieth wedding anniversary with an open house at their farm. Just two years later, Grandpa Roy Crawford died in June 1960 at the age of 78, and was buried in the Luray Cemetery. Dad had already received his share of Grandpa's land, and the remaining Crawford land went to Grandma and Pauline with the stipulation that it would all go to Pauline upon Grandma's death. Grandpa had also owned the house in Luray that Mother and I had lived in during the war, and his will left this house jointly to Pauline and Dad.

After Grandpa's death, Grandma moved into that house, where she could be close to the grocery store, drug store, and other conveniences, and from that point on there were no longer any families living on section 35.

60. Albina and Roy Crawford, fiftieth wedding anniversary. (1958)

In the early 1960s, Lloyd and Bertha Maude Beatty bought the farmhouse Grandpa built, the place where he and Grandma lived most of their lives. The Beattys moved that house to their farmstead in section 34, where it became their new residence. Charles Shaffer purchased the barn on Grandpa and Grandma's farmstead, and moved it to his farm about three miles south of Waldo. The departure of those two main buildings left only the shelter belt, the well, and a few of the minor farm buildings as a reminder of what an active place the Roy Crawford farmstead had once been.

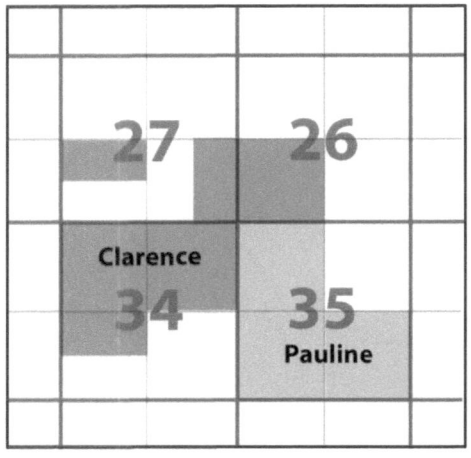

61. Crawford land in 1963.

In 1962, Dad and Pauline sold all their land in section 32 (240 acres each) to Lyle Trapp. At the same time, Dad purchased 240 acres from Ivan and Margaret King (SW¼, 26-11S-13W and E½ SE¼, 27-11S-13W). This consisted of 80 acres of pasture in section 27 and 160 acres of mostly cropland in section 26. All this King land was adjacent to the rest of the Crawford land, and included the original King farmstead in section 26 where Ivan, Margaret, Newt, and Kay had lived. The net effect of these transactions on Dad's farming operation was that he still owned 720 acres, but now less of it was pasture and more of it was cropland. It also consolidated his farming operations by eliminating the need to move machinery back and forth over the roads to section 32. Our family no longer owned the pond in section 32 where I had spent many hours fishing, but that didn't matter since none of the members of our family left in the area wanted to fish there.

In 1962, Grandma Crawford was diagnosed with Hodgkin's lymphoma and needed to be closer to medical support, so Dad and Mother bought a house in Russell and the family, including Grandma, moved into Russell in the summer of 1962.

This move meant that Candace could finish her remaining high schooling in Russell, where the educational opportunities were broader than those offered at Luray—another factor behind the decision to move. Dad kept all the farming equipment at the farm, but he rented our farm house to the Bob Rose family for a nominal fee just so the house would remain occupied and not immediately fall into disrepair.

Grandma Albina (Hampl) Crawford died in June 1963; she is buried beside Roy in the Luray Cemetery. As Grandpa's will stipulated, Pauline then inherited all the land Grandpa and Grandma Crawford had owned in section 35.

Pauline rented all her section 35 land to Dad, and he continued to farm it along with his own land after that. On February 15, 1965, Dad as the operator and Pauline as the owner agreed to a soil conservation plan covering Pauline's land in section 35. Dad designed the plan in conjunction with the Russell County Soil Conservation District, assisted by the United States Department of Agriculture Soil Conservation Service. This plan involved the construction of waterways and terraces on that land, and Dad made sure that happened.

Although short-term needs motivated their move, Dad and Mother continued to reside in their Russell house for the remainder of their lives. Soon after the move, Dad joined the Rotary Club in Russell and Mother joined a Federated Club there, so they quickly integrated into the Russell social fabric. Dad continued to operate the farm after the move to Russell, taking a can of tuna and some crackers for his lunch and commuting the 17 miles each way almost daily during the remainder of the summer and through the fall.

This commute (and the tuna and cracker lunches) would continue through the busy farming season, spring through fall, for the next twenty years. To simplify the farming, Dad disposed of all the livestock, so that he only needed to take care of the tilling, fertilizing, planting, and harvesting of the crops (generally just wheat and milo).

Mother accepted a job teaching English in the Russell High School, starting in the fall of 1963, and she continued to teach there for the next ten years.

Dad's simplification of his farming operation provided an unexpected benefit. It left Dad with more free time and enough schedule flexibility so that he and Mother could travel frequently even with the demands of her school schedule. Over the next decades, they traveled extensively, visiting parts of Europe, the Middle East, Northern Africa, South America, and Australia.

In the late 1950s and early 1960s, Dad was chair of the Russell County Extension Council, an advisory group to the county extension service. This council helped identify educational needs, develop extension programs, and assist with program evaluations in the areas of agriculture, family and consumer sciences, 4-H youth activities, and economic development.

Governor George Docking appointed Dad to the Kansas State Board of Education for a three-year term beginning January 1, 1960, and Dad remained on that board when the family moved to Russell. His appointment to a second three-year term retained him on the board through 1965, serving the maximum number of terms allowed. He functioned as Chairman of the Board during two of his six years as a board member. During the years he was on the board, it reorganized the State Department of Education and initiated the state program for area vocational training, which included a new rehabilitation center in Salina.

Candace graduated from Russell High School in the spring of 1964, and enrolled at the University of Kansas in Lawrence in the fall of that year. She graduated from KU with a degree in English

Education in the spring of 1969. After a few years of working in Princeton, New Jersey, Candace went back to school, attending the University of Western Michigan at Kalamazoo where she acquired a Master's Degree in Occupational Therapy. Following her graduation, she began her career as an occupational therapist working in Los Angeles.

She married Tim Keiderling on September 4, 1976, and they settled in Western Springs, Illinois, as Tim assumed a faculty position in the Chemistry Department at the University of Illinois, Chicago Circle. They later moved into Chicago. Occupational therapists were in high demand and Candace could take her pick among several different places to work in Chicago, including the University of Illinois Medical Center and Northwestern Memorial Hospital. She also taught courses in occupational therapy at the University of Illinois at the Medical Center, Chicago. In her spare time, she took courses in creative writing at the University of Illinois, Chicago, leading to a Master's degree in that field. She too had chosen a path that did not lead back to the farm.

In the late 1960s, Dad traded in his Case tractor on a new John Deere diesel tractor, but otherwise he continued to farm with most of the same equipment that he had accumulated through the 1940s, 1950s, and early 1960s. That John Deere tractor was the last major piece of farm equipment that he purchased. In the late 1970s, Dad decided to stop farming Pauline's land, and Pauline then rented her section 35 property to Eldon Hampl to farm. Eldon continued to farm this land for several years.

Dad had focused on conservation throughout his years of farming, and in 1976 he received the annual Goodyear Award for his "outstanding soil, water, and wildlife conservation practices and for his constant promotion of conservation." He had previously received the Kansas Bankers' Soil Conservation Award recognizing his "outstanding progress in practicing conservation" on his farm.

62. Clarence Crawford receiving the annual Goodyear Award
from Goodyear representative Dick Howe of Salina and
Goodyear dealer Roy Cable of Russell. (1976)

In 1988, Dad once again assumed the role as the farm operator for Pauline's section 35 land, and in July 1988, Dad as the operator and Pauline as the owner agreed to a revised soil conservation plan covering that land. As with the other conservation plans, they designed this plan in conjunction with the Russell County Soil

Conservation District with the assistance of the United States Department of Agriculture Soil Conservation Service. This plan included completion of the remainder of the terraces on the section 35 land. Once this plan was in place, Delmar Hampl took over the farming of Pauline's land.

63. The family at the fiftieth wedding anniversary of Clarence and Gladys Crawford in 1989. **Back Row:** Tim Keiderling, Gladys Crawford, Clarence Crawford, Kent Crawford. **Front Row:** Candace Crawford, Lara Crawford, Michael Keiderling, Clare Crawford, Charlotte Crawford.

By the late 1980s, Dad was in his early 70s. He had largely retired from operating his farm, and was leasing it all to Delmar. However, Dad continued to maintain an active interest, and was often at the farm observing the farming operations or working on a project of his own. One of his major goals as a farmer had been to complete the system of waterways and terraces on all the Crawford land, leaving that land well protected from erosion and in a much better condition than when he had first begun farming. He had seen that goal completed by the time he retired from operating the farm, having lived up to his motto about the need to be good to the land.

While he was alive, Grandpa Crawford had expressed the wish that the Crawford family continue to own his land after his death. Pauline intended to honor this wish, and since she had no children of her own (Dalene and Robert were her stepchildren), she decided to transfer her holdings of Grandpa's land in section 35 to Candace's family and to my family. The net result was that after the transaction was complete in 2000, my family and Candace's family each owned a 50 percent interest in that land, and Delmar Hampl continued to farm it for us.

Pauline was a remarkable person, and this generous act was just one example. In 2000, the community of St. John recognized her many contributions as well—the Xi Zeta Eta sorority named her "Lady of the Year" for her many years of community service in the St. John area.

Disappearance of the Rural Population

In the early 1970s, the Soviet Union decided to purchase large quantities of wheat and corn in international markets to improve the diets of Soviet citizens. At the same time, the US withdrew from the gold standard that tied the U.S. dollar to gold at the rate of $35 per ounce, causing the US dollar to decline in value relative to many other currencies and making US agricultural commodities more attractively priced to foreign buyers. Both actions led to a huge expansion of agricultural exports, which raised commodity prices and contributed to an upward spiral in land prices and an intense period of investment in machinery, farm buildings, and rural housing. In many areas farmland prices increased at about twice the general inflation rate from 1972 to 1979.

The price of wheat jumped due to the increased demand, and remained relatively high for the rest of the decade, while wheat yields continued their general upward trends. Harvests between 1974 and 1979 reached record levels, and much of that increase in production was from farmers who had bought additional land and newer machinery with borrowed money. In 1962, total US farm debt was $60 billion; by 1983, farm debt had skyrocketed to $216 billion. Land and machinery prices reached new highs in the production rush of the 1970s. Export markets increased significantly during the 1970s; farm exports in 1960 totaled only $6 billion, but by 1979, they had quintupled to over $32 billion, and that year the US exported a quarter of all crops grown.

Then in 1980, President Jimmy Carter placed an embargo on sales of grain to Russia following the Russian invasion of Afghanistan, causing exports to stagnate and to begin decreasing. During the early 1980s, inflation was rampant across the US, reaching as high as 17-18 percent. Since wheat prices remained constant or dropped during this same period, farm incomes could not keep up with the cost of living, and this created a financial squeeze on the farmers. Land prices fell by 30-50 percent and farmers who had borrowed money with high land prices as their collateral often couldn't find new loans, even for operating expenses. Many farmers who had bought land in the 1970s began to default on their debts and by mid-decade many farm businesses were in liquidation and dozens of agricultural banks were failing.

As exports and market prices slumped, the government took title to hundreds of millions of bushels of wheat and corn forfeited through nonrecourse loans. The government paid the farmers to store the forfeited grain, because there were not enough other storage facilities. Treasury outlays for the non-recourse loans and other subsidies amounted to $26 billion in 1986, and together with an easing of the slump in exports, this massive bailout halted the downward slide in farm incomes and farmland values by the late 1980s.

The financial squeeze of the 1980s drove many Kansas farmers off their farms. However, throughout the 1970s and 1980s, Dad had continued the Crawford philosophy of not buying anything unless he could justify it and could pay cash for it. Hence, the Crawford farm operation did not suffer the same debt squeeze experienced by many other farmers during that period.

By the end of the 1960s, only the Lloyd Beatty family and the Bob Rose family were living on section 34, which had at one time supported four homesteading farm families. Dad was still farming all his land at that time and he continued to keep his farm equipment in our farmyard and machine shed on section 34. No families were living on section 35 where Grandpa and Grandma once lived or on the southeast quarter of section 26 where the Ivan Kings and earlier the Harry Kings had lived. Most of these lands were now part of the Crawford farm.

Bertha Maude and Lloyd Beatty passed away in the 1980s, and their son Tom continued to live on their farm for a while. However, no one was living on section 34 by the end of the century.

By the beginning of the 1960s, the Union Pacific Plainville Branch rail line serving Luray and Waldo had discontinued passenger service, but the railroad still provided regular freight service along this line, and still transported the wheat and other grains accumulated in the local elevators. This service continued until 1993, when flooding on the Saline River severely damaged the tracks and bridges along parts of that branch line. The Union Pacific decided it would be too expensive to repair that damage, and instead discontinued all rail service on this branch line. Trucks took over delivery of the supplies for the remaining merchants and others in Luray and Waldo areas. Trucks now also transported the wheat, milo, and any other farm products purchased by or sold at the Luray Co-op elevator.

The Luray population dropped from 328 in 1960 to 303 in 1970. However, the high school population increased during this decade due to the post-war baby boom, growing from 115 seniors graduating during the decade of the 1950s to 139 graduating during the decade of the 1960s. The population of Waldo dropped from 178 in 1960 to 123 in 1970, and Waldo lost its high school during this decade. The class of 1964 contained the last seniors to graduate from Waldo High School, which closed in 1964 when Waldo High School consolidated with Paradise High School; Waldo high school students then began attending the school in Paradise, a small town in the northwest corner of Russell County.

In 1974, the Paradise, Natoma, and Waldo school districts combined to form unified school district USD 399 Paradise-Natoma-Waldo, with elementary and high schools both located in Natoma. Paradise, six miles west of Waldo, was in Russell County; Natoma, eight miles northwest of Paradise, was in Osborne County. At that point, some of the Waldo students opted to attend Luray High School and/or Luray Grade School instead, since Waldo was much closer to Luray than to Natoma.

The declining high school population finally forced Luray High School to switch from 11-man football to 8-man football in the 1970s. True to Luray High School football tradition, they managed to go out in a burst of glory, winning the Kansas State High School 8-Man Football championship in 1975. The class of 1977 was the last class to graduate from Luray High School.

In 1977, the Luray and Lucas school districts consolidated as USD 407 Russell County Lucas-Luray Schools, with the elementary school located in Luray and the high school located in Lucas. This arrangement persisted until 2010 when Lucas-Luray combined with Sylvan Unified Schools to form USD 299 Sylvan-Lucas Unified, which has attendance centers in Lucas and Sylvan Grove but there is no longer any public school in Luray. Sylvan Grove, located in Lincoln County, is about eight miles southeast of Lucas and eighteen miles southeast of Luray.

One of Luray's most steadfast champions, John A. O'Leary Jr., the Peoples State Bank president, died in 1989. In February 1992, the People's State Bank in Luray became a branch of the UMB banking system. Eventually the UMB bank decided the Luray Branch was too small for its system, and the Farmers Bank of Osborne bought the Luray bank branch from UMB, effective July 2011. The Farmers Bank of Osborne merged with The Peoples Bank (not related to the Peoples State Bank), and began operating under that name in 2020. Despite the name changes, business as usual continued at the Luray bank, but with the loss of John, the bank was no longer such a strong positive community proponent.

The population of Luray continued its decline, falling to 295 in 1980, and to 166 by 2020 as the farming bust of the 1980s took its toll. The population of Waldo dropped precipitously during this period, falling from 123 in 1970, to 75 in 1980, and to 30 in 2020. Luray continued to function as a town, with a bank, a grain elevator, a post office, and a few other businesses; but Waldo, with a population of only 30, had largely disappeared as a town by this time.

This local population decline was symptomatic of the general decline in farm population in Russell County and beyond. Only 156 people lived in Russell County in 1870, but there were over 7,300 ten years later in 1880. Population rose steadily to a peak of nearly 13,500 in 1940, driven partly by a boom that followed the discovery of oil in the county in 1923, but after a steady decline the county population was back to about 6,700 in 2020. Furthermore, 4,400 of the Russell County residents were in the city of Russell and another 1,100 were in the other towns in the county by then. Subtracting the population of Russell and the other towns from that of Russell County provides an estimate of the number of people living on farms in the county. This farm population reached a peak of about 7,000 in 1910 through 1920, but then began to fall steadily to about 1,200 (an average of a little more than one person per section) in 2020.

Becoming a Century Farm

In the United States or Canada, the Century Farm designation is an award presented to a farm or ranch based on documentation that a single family has continuously owned the farm for 100 years or more. In Kansas, the Kansas Farm Bureau grants this recognition, and reserves the Century Farm designation for farms of at least 80 acres. The Farm Bureau initiated the Kansas Century Farm program in 2000, and by 2020, the Farm Bureau had recognized nearly 3,000 Century farms in Kansas, 59 of which were in Russell County.

The Crawford farm had already been in the family for more than 100 years by the time the Century Farm program started in Kansas, and Dad applied for the Century Farm designation shortly thereafter. He received this recognition for the Crawford farm on November 16, 2001, and was justifiably proud to have been a part of the multi-generational accomplishment represented by this designation.

The history of the Crawfords and their farm provides some understanding of just what the appellation Century Farm represents. Although details of this history are specific to the Crawford family, many other multi-generational Kansas farm families, including those with Century Farms, have followed quite similar paths.

64. Century Farm Award.

To reach 100 years of family farm ownership is virtually impossible without the contributions of at least two and often more generations of family members. In the case of the Crawfords, the first generation on the farm was Harmon and Candus, who homesteaded to establish the farm and survived many hardships while managing to turn that homestead into a farm that could support their family. They acquired additional land and took advantage of the advances in laborsaving farm implements using teams of horses rather than manpower to provide the power necessary to farm this additional acreage.

65. Dad receiving Century Farm Award from Randyl Becker, Russell County Farm Bureau President. (2001)

Roy and Albina followed them, expanding the farm with the purchase of more acreage as they negotiated the transition from horse-powered farming to farming with gasoline powered tractors and trucks. Astute management and excellent understanding and application of the best farming practices enabled them to prosper despite the low prices for wheat throughout the 1920s, and the hardships imposed by the Depression and the Dust Bowl in the 1930s.

Clarence and Gladys took over the farming after the end of WWII. Clarence applied himself toward improving the Crawford farmland, introducing terraces, contour farming, use of chemical fertilizers, new strains of wheat, and the inclusion of milo as part of the regular crop rotation. He also negotiated land purchases that made the land more convenient to farm, and carefully navigated the myriad of government farm programs that required significant changes to farm management practices. His stewardship enabled the Crawford farm to continue to prosper through a time when many farmers were finding it difficult making ends meet, and many were leaving the farms to find work elsewhere.

This lengthy multi-generational effort required perseverance, great respect for and love of the land, together with the satisfaction derived from working to coax the land to yield its bounty. Those traits, present in the original homesteaders, carried on to the two succeeding generations as well. Harmon, Roy, and Clarence all loved farming and had a certain reverence for the land itself and what it could produce. They each had good business sense, and were conservative in their business practices. They all felt that it was better to do without something than to go into debt to buy it, with the only possible exception of going into debt if necessary to buy more land. All three of them recognized the importance of employing new technologies and of applying the lessons learned from research on best farming practices and new strains of wheat, but only employed those technologies or applied those lessons when they deemed these new practices to be cost effective. Each of their families provided the steadfast support they needed to prosper in their endeavors. These traits were responsible for the successful operation of the Crawford farm throughout the three generations of Crawford owners and operators, allowing that farm to reach Century Farm status.

Our family, like countless other Kansas families, represents a mix of ethnicities brought by the early settlers as well as by those who married into the family. Kansas attracted settlers from many different backgrounds, each for their own reasons.

These different ethnic groups brought their own experiences, cultural traditions, technical and general knowledge, and new techniques for solving practical problems. This "melting-pot" diversity was largely beneficial, and the succeeding generations were able to adapt portions of each of these cultures to create and perpetuate their own traditions, building on the strengths brought with each ethnic group. This diverse background likely also played a role, albeit difficult to ascertain, in the Crawford farm successfully reaching Century Farm status.

End of an Era?

Our family witnessed and participated in the rise and fall of the rural population in Kansas during the more than 140 years the Crawford farm has been operating. When Harmon and Candus first settled on their homestead, much of Kansas was still open and there were no nearby towns. By the early 1880s, families occupied all four quarters of the section on which the Crawfords initially settled. Other settlers inhabited farms on the nearby sections in the Amherst area, which quickly became a thriving community. Nearly every farm had neighbors within a half mile in each of several different directions. By 1890, the new towns of Luray and Waldo served by the branch railroad line were becoming the primary shopping centers for this community.

The populations and the fortunes of those towns and that community mirrored the changes in the farm population as the towns and the community progressed through the 1890s and into the twentieth century. Luray and Waldo grew rapidly, reaching 400 and 200 inhabitants respectively by 1915, and they became the main trading centers for the farmers from the Amherst community. Both towns continued to thrive, maintaining their populations at or above those levels through the 1930s.

However, as the farms started getting larger and the farm population began its steady decrease in the 1930s and 1940s, the populations of those towns began to decrease as well. The decline in the farm population also had a profound effect on the school systems in Luray and Waldo, and school consolidations eventually left both towns without schools of their own. Those schools had once served as focal points of the social and community activities in Luray and Waldo, and their loss removed most of the remaining reasons for the local farm populations to identify with Luray or Waldo as their hometowns. By the early twenty-first century, the whole area had reverted to a place with few inhabitants and few viable small towns, eerily reminiscent of the status when the first settlers were arriving.

Ultimately my parents found that the Luray and Waldo communities could no longer supply what they needed and they moved to Russell, choosing a new home that required Dad to commute to farm the Crawford land. The Crawford farm now consists of about 1,200 acres of land in four adjacent sections. No Crawfords or any others live on any of this land, but one can find the remnants of at least four different abandoned farmsteads there that serve as reminders of how different this area once had been.

More has disappeared than just the population — a community has disappeared as well. In the 1950s when I lived in the area, Luray and Waldo each had both a high school and a grade school. Much of the community identified with those schools, and community life revolved around those schools. Parents and other community members, often alumni of those same schools, supported school sports, musical or dramatic performances, and other school functions. Parents also got to know one another through involvement with their children's school friends. This connection with the schools led people to do their shopping in Luray or Waldo and to participate in community organizations such as the Lions Club and the American Legion, bridge clubs or other social clubs, and church-affiliated organizations, and this also fostered the feeling of community. Saturday nights spent shopping and visiting in town reinforced this community connection, as did drop-in visits with neighbors.

As people continued to leave the farms, the population remaining in the area was insufficient to support some of the businesses in the town and the loss of those businesses also meant a loss of the jobs they provided, further contributing to the population drain. As the population of the area decreased school enrollments fell, and ultimately Luray and Waldo each lost all their schools. Now, instead of being at most a few miles from the school, students from the Luray and Waldo areas take buses elsewhere. Luray students go 10 or more miles to grade school in Lucas, or up to 30 miles to high school in Sylvan Grove; Waldo students travel about 15 miles to Natoma. People now must go to Lucas or Russell or Natoma for shopping that they once would have done in Luray or Waldo, and travel to Lucas or Sylvan Grove or Natoma if they want to attend school functions.

Gone are the carefree Saturday nights in town. Gone are the days when the spirit of the community rose and fell with the fortunes of the local high school football team. Gone is the sense of community that helped sustain farmers and townspeople through difficult times. With the diminished shopping opportunities and the absence of schools, the main connections between the rural areas and the town of Luray now are the Co-op elevator and the branch bank, not enough to rekindle whatever community spirit remains. Even less connection to Waldo remains, as there is little or no business activity there.

Luray is still trying, though. For more than 50 years they have been hosting an annual Friendship Day with free food, activities for children, and live entertainment. The Methodist men of Luray have continued to host an annual fish fry for even longer than that, and this still draws large crowds. The association of Luray high school alumni has also hosted an annual alumni banquet for many years, now expanded to include alumni of the consolidated Lucas-Luray high school. The success of these annual functions is a testimony to the attachment the school alumni and other former members of the community still have to the schools and community of their memory, but although these efforts may have helped slow the steady decline of the town, they have not translated into a revitalization of the Luray community.

In the 1960s, circumstances induced Dad and Mother to move off the farm and into Russell, making them a part of the general exodus of population from the Amherst and Luray communities. However, Dad continued farming for another twenty years after he and Mother moved to Russell, commuting back and forth between their house in Russell and the farm. Dad engaged in some land transactions to simplify his commute. These transactions were not to expand the farm, but rather consolidate it so he could farm it more efficiently. Dad and Mother followed the Crawford practice of never going into debt, and this stood them in good stead throughout the wildly fluctuating economic conditions of the 1970s and 1980s.

Dad continued to promote conservation activities, working closely with the Farm Service Agency. He received recognition for his persistent and effective promotion of conservation. However, Dad and Mother also found time to pursue their support of education, one of their other interests. Mother went back to teaching, serving on the faculty of Russell High School, and Dad served on the State Board of Education, traveling to Topeka for the board meetings. They also found time to travel extensively—a continuation of a family tradition.

When Dad finally retired from active farming in the late 1980s, he and Mother had certainly lived up to the motto: "This land has been very good to our family, and we need to be good to it!" Not only did they improve the land, but they also improved the communities in which they lived.

Mother and Dad celebrated their sixtieth wedding anniversary in 1999, and three years later, Mother died in October 2002, at the age of 92. Dad died in May 2008, also at age 92, and he lies next to Mother in the Luray Cemetery. Their tenure as owners of the Crawford farm had spanned more than 50 years.

Candace and I are the only remaining Crawfords who lived on and participated in the operation of the Crawford farm. We are also the only remaining Crawfords who can remember when Luray, Waldo, and the Amherst community were all well-populated vibrant communities with which we could readily identify—a much different place than what rural Kansas has now become.

After graduating from college, Candace and I each followed paths that led away from farming, so none of the fourth generation of Kansas Crawfords has been directly involved in the Crawford farm operations since that time. Ownership of the Crawford farm still resides in the Crawford family, but there is no expectation that any of our descendants will ever farm there. Kevin Hampl, who took over from his father, Delmar Hampl, is currently handling farming operations on that land, so care of the land is still in the hands of a family member, albeit a member of our extended family. Kevin and one full-time co-worker not only farm all the Crawford land, they also farm Hampl land and other leased farmland— symptomatic of the scale now typical of Kansas farm operations.

66. Candus' frame house, the house I grew up in, as it stood empty in 2008.

67. Same site in 2008 after burial of the house and some of the trees. The unoccupied King farmstead is now visible in the background.

By 2008 the farmhouse we had lived in had been standing empty on the Crawford homestead quarter for some time. Empty farmhouses typically deteriorate rapidly due to weather, vandals, raccoons, and other pests, and that was why my folks rented their farmhouse to the Bob Rose family for a nominal fee. However, time took its toll, and after that house had survived nearly a century, it reached such a state of disrepair that the Roses moved out, leaving the house abandoned. We buried the remains of that house in 2008; nearly 100 years after Candus had built it. The land where it stood has now reverted to grassland, leaving few signs of the families who had lived there through the years. At the time of this writing, it is not clear what the future holds, and how much longer the Crawford family relationship to this land will continue.

I found growing up on that farm and participating in the Amherst and Luray communities were wonderful experiences, but there is no going back. The farmhouse that was my home is gone, my grandparents' house and farmstead where I roamed freely during my childhood years are gone, and the schools that I attended are gone. Most of the people are gone from the farms, and the town that I considered to be my home town is almost gone. The majestic panorama of open space provided by the High Plains remains, now dotted with artifacts providing hints that many more people once lived here. The wheat and milo crops continue to grow in the fields and the scorching hot summers and the incessant winds also remain, but there is little else left that resembles the place of my youth.

Yet this was my home and I still have fond memories of the way it once was, a vibrant community populated by numerous farm families and supported by the businesses and school systems of several small towns.

In these remembrances, I can once again walk the two blocks of Main Street with my friends and hang out at the drugstore in Luray, and can join that successful football team for one more dream season. I can also ride round and round the fields, tilling the ground while the tractor mostly steers itself as I contemplate the majesty of the wide-open spaces and the far horizons. Better yet, I can navigate the combine through the field of ripe billowing waves of wheat, mindful of the fact that a golden stream of grain is descending from the spout into the bin behind me.

I revisit those memories from time to time, and I will forever be grateful for the privilege of having spent my formative years with my family on this farm and in this community, participating in this place and in this era central to so much of my family's history. I would not trade this experience for any other alternative I can imagine.

What Next?

The Crawford attachment to this land and this area will end at some point, but the land remains. It still grows the huge crops of wheat that have made Kansas famous as "the breadbasket of the world," although now it requires only a few farmers to produce those crops. The world will continue to need food, so it is likely that most of this land will remain in production for the near future. However, the nature of farming on the High Plains has changed drastically through the past century and a half, and it is likely to continue changing, perhaps in unforeseen ways, as the few remaining farmers adapt to these and other new challenges.

During the more than a century that the Crawfords have been farming in Kansas, operation of the Crawford farm underwent numerous transformations, as did farming in western Kansas in general. The initial settlers established nearly self-sufficient sustainable small farms with a diversity of crops and livestock. Operation of these farms was exceedingly labor intensive.

However, the mechanization of most of the farming operations proceeded at a rapid pace, with horses and later tractors providing the power for these operations. These changes vastly increased the ability to work a larger farm with one or a few persons. Horses and tractors required cash for their purchase, and this led farmers to focus on cash crops, quickly settling on wheat as most suitable for western Kansas.

The technological advances and the successful use of chemicals and better seed to increase the productivity of the land, mean that now farmers can produce much more wheat and milo than they produced at the peak of the Kansas farm population, even though there are now far fewer farmers.

Once all these farms were producing wheat, wheat became a commodity subject to supply and demand economics largely out of the control of the farmers, and it became nearly impossible for a small farm to produce enough of this commodity to survive. Kansas commodity farming of wheat and milo is heavily dependent on modern machinery and on extensive use of fertilizer, pesticides, and herbicides. A few giant agribusiness corporations now dominate these supplies, and consequently machinery and chemical costs have risen dramatically.

As an example, Clarence's Massey-Harris combine cost about $5,000 in 1948, while the minimum cost for a basic combine in 2010 was about $250,000. The general cost of living increased by a factor of seven during that period, in contrast to the factor of 50 increase in the dollar cost of a combine.

Looking at this data another way, the price of wheat in 1948 was about $2 per bushel, of which on average perhaps 10-15 percent would be profit. If one assumes a 15 percent profit, then in relevant terms the 1948 combine would cost the profit from about 17,000 bushels of wheat. A good yield at that time would have been about 20 bushels per acre, so one year's profit from 850 acres would pay for the combine in 1948.

In 2010, the price of wheat was about $7 per bushel, so in equivalent relevant terms with the same profit assumptions the 2010 combine would cost the profit from at least 240,000 bushels of wheat or 14 times as much as the 1948 combine. Even with a yield of 60 bushels per acre, relatively high for 2010, paying for this combine would still require all of one year's profit from at least 4,000 acres, about five times as many acres as required in 1948. This sets the scale for the size of farm required to retain viability in 2010.

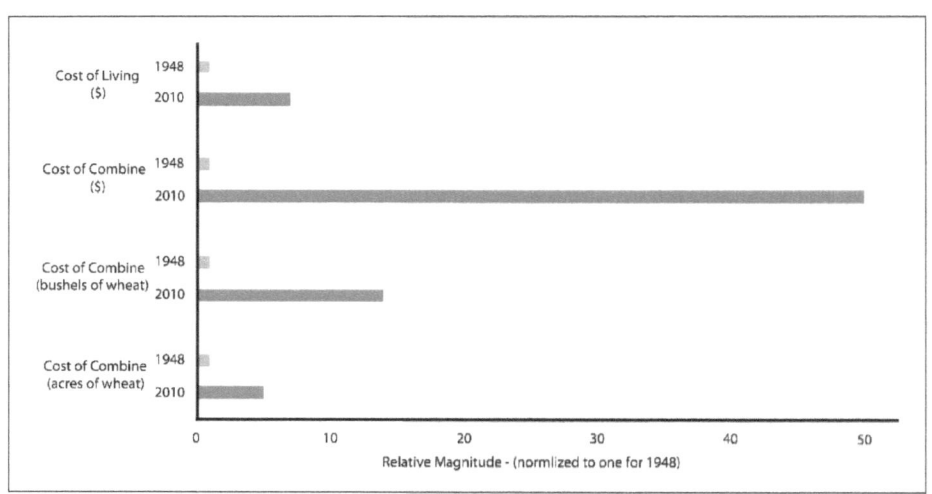

68. Combine cost comparison—2010 vs. 1948.

Of course, the 2010 combine had many more capabilities than the 1948 combine, but it was now priced out of the reach of all but the largest scale farms. Figure 68 summarizes these comparisons.

The astronomical increase in the price of a combine represents one of the extreme cases, but other types of farm machinery show similar effects as well.

A yield of 60 bushels per acre at $7 per bushel produces a gross annual income of $1,680,000 on 4,000 acres of wheat, so large-scale farming is big business. The costs of the machinery, not only combines but tractors, trucks, tilling tools, chemical application rigs, etc., are huge. Costs of chemicals are high and rising, and there are other production costs such as fuel, insurance, and taxes. Small changes in the price of wheat or milo, or changes in yield (e.g., a bad year due to drought), both of which are largely out of the control of an individual farmer, can make big swings in the gross income. Since production costs and gross income are both very large, a slight swing in either can have a huge effect on net income. Big farming means big profits are possible, but large losses or even bankruptcy loom as very real threats.

In addition to the rapidly growing costs of machinery and chemicals, Kansas farmers now must face the prospect of dramatic adjustments necessitated by climate changes and by the ongoing advances in the automation of farming operations.

Climate change models predict increasingly more violent weather separated by lengthening periods of drought, and this will likely increase the difficulty of obtaining good yields of wheat and milo with dryland farming such as that on the Crawford farm. It may even force an acceleration of the return of some of the dryland cropland to native prairie.

Irrigated farmland faces the rapid depletion of the Ogallala Aquifer (also known as the High Plains Aquifer), with many, if not most, of the Kansas irrigation wells expected to run dry within the next few decades. Yields of wheat and milo from currently irrigated land will drop by nearly a factor of two when those wells stop pumping. The large western Kansas feedlots that depend on water from the aquifer may eventually need to curtail their operation, possibly decreasing the demand for Kansas milo as well.

Another revolution in farming practices, often called the "digital revolution" or "automated farming" or "smart farming," has been underway for some time. This revolution encompasses a multitude of changes, some of which are already available and in use by many farmers. Implement-mounted computers and artificial intelligence (AI) already enable some currently-available implements to adjust themselves to optimal configurations as the terrain and conditions encountered by the implement change; such enhancements are part of what drives the high cost of a modern combine. Drones now provide an easy means of detailed localized aerial assessment of field and crop conditions. AI-based business software for farmers helps optimize the allocation and application of resources such as fertilizer and water.

However, this is just the beginning. On the horizon is the use of multiple robots to replace heavy tractors and other large equipment, providing fine-scale optimization of tilling, planting, fertilizing, and perhaps even harvesting, while avoiding the soil compaction associated with the heavy equipment. It does not take a large stretch of the imagination to visualize a future farm with all the farming operations carried out by an army of robots directed by the AI-driven business software. In such a scenario, technicians to keep the robots and software in good repair and functioning smoothly might replace the farm workers. It might even be possible eventually to perform all the farming operations remotely, in which case the farm operator need not live nearby but could instead reside in another state or even in another country. At that point one could reasonably ask "What does it mean to be a farmer?" or even "Are farmers really needed, or will the agribusinesses or other entities take over this factory-like operation?"

It will be some time before automation totally takes over farming. In the nearer term, smart farming techniques can aid in mitigating some of the effects of climate change on farm production and profitability. However, the planting of alternative crops or even engaging in alternative uses of the land are also options to consider instead of or in addition to the smart farming techniques.

In recent years foreign production of high-quality hard red winter wheat has increased significantly, keeping wheat prices in the US low and making it difficult for Kansas farmers to grow wheat profitably. The acreage planted to wheat in Kansas has been dropping, with some farmers turning to other crops such as corn, grain sorghum, soybeans, sunflowers, forage, or even cotton. Grain sorghum continues to replace some of the wheat acreage in western Kansas, and since the development of drought-resistant strains of corn, corn is also becoming a practical alternative even on the high dry plains of western Kansas. If new strains of some of these other crops can thrive on lower moisture input, perhaps they will take over some of the western Kansas farm acreage in the future as well.

Kansas has long been famous for its hard red winter wheat. However, the agricultural experiment stations at Hays and elsewhere around the country have developed hard white winter wheat to the point where it is starting to make inroads on the long dominant hard red winter wheat. So far, there is little or no foreign competition to drive the prices down for the hard white wheat, but there is little US infrastructure to handle this wheat separately from the hard red wheat and that limits the market for this wheat. Development of the necessary additional grain-handling infrastructure to process the white wheat separately from the red would likely provide a significant boost to the importance of hard white winter wheat as a Kansas crop.

More than ten years of development have gone into producing perennial wheat strains. So far, these developments have shown some successes, but the yields are still well below those for annual wheat and the strains need to be replanted every two or three years. Nevertheless, these perennial strains require less tillage and exhibit deep root growth, and this helps with soil improvement and provides more carbon sequestration in the soil, so there is a strong incentive to continue development of the perennial wheat strains. Improvement of the perennial strains to the point where they are competitive with the present annual wheat varieties is the goal, but that outcome remains far from certain.

Another alternative is to turn cropland formerly used for wheat into grassland that can be adapted for grazing or to support wildlife. The Department of Agriculture's Conservation Reserve Program (CRP) pays farmers to convert cropland that qualifies for the program (highly erodible or otherwise fragile land) into grassland, removing that land from agricultural production and planting species that will improve environmental health and quality. These payments continue for a contracted period, typically 10-15 years. The goal is to reestablish year-around ground cover that will help improve water quality, prevent soil erosion, and reduce loss of wildlife habitat.

There is a wealth of evidence showing that such a move can result in significant improvement of the soil quality over time, especially if the cropland soil was of relatively poor quality. Some farmers have chosen to leave the land in grass even after those contracts expire. They have found that with proper management, using the land for grazing can be more profitable than using it for growing wheat. However, CRP targets only the most environmentally sensitive portions of farms. Enrollment of whole farms or large tracts of productive farmland would potentially distort the market, signaling competitors to plant more acres. This could result in conversion of rain forests or other environmentally fragile land elsewhere into cropland.

Some groups have touted the conversion of cropland to grassland used for grazing as a means of sequestering huge amounts of atmospheric carbon, and hence as an important tool for fighting greenhouse-gas-induced global warming. However, the evidence so far does not support most of these claims. Converting poor cropland into grassland may sequester some additional atmospheric carbon, but the effects are not large and appear to saturate a few years after the conversion—not enough to significantly impact global warming. Nevertheless, the Department of Agriculture has recently announced that it is beginning a new program to sample, measure, and monitor soil carbon on CRP acres. This program should eventually provide much-needed data to clarify these effects and perhaps help direct future land-use decisions.

Recreational use is another potential application of Kansas cropland or grassland. It can include hunting, fishing, camping, hiking, weekend retreats, and other activities that get one outside and in contact with nature. Sale or lease of land for recreational use is becoming more widespread and is another potential source of income for farmers. Russell County Kansas is a prime area for hunting, especially for pheasants which thrive in a mixed cropland and grassland habitat.

Time will tell which, if any, of these alternatives will become important for High Plains farmers in the future. Climate change will likely play a role in such decisions, as will public policy, implemented through the government farm programs. Further technological advances, including advances in automation, will influence future farming decisions as well. The only certainty is that farming on the High Plains of Kansas will continue to change, with the form and extent of these changes not yet determined and perhaps not yet even imagined.

Epilogue

The house and barn had been gone from the Roy Crawford farmstead in section 35 for many years, but the yard and the lush stand of trees forming the shelter belt that surrounded the yard remained. However, all that changed on the day the fire came.

Unseasonably warm temperatures and weeks without rain left much of Kansas at high risk for fires in December, 2021. An extraordinary storm on December 15 generated wind speeds approaching 100 miles per hour, snapping electrical power poles and lines which then started numerous prairie fires. Fires in 11 counties from Russell County west burned nearly 400,000 acres, with about 365,850 acres in a fire that stretched across four counties: Ellis, Russell, Osborne, and Rooks. In Russell County alone, the fires burned between 200 and 250 square miles and caused millions of dollars in damage. The fires destroyed several homes, outbuildings and other structures and killed many cattle and horses.

One of these fires reached the Amherst community, and did damage to the Crawford holdings there. The wind blew straight from the west—lined up with the roadside ditches and fence lines. The weeds and grass growing in those fence lines made those lines into efficient spreaders of prairie fires, another reason to remove any unneeded fences. Most of the ditches and fence rows in those areas burned, and the barbed wire and stone posts in those areas sustained damage. Some of the posts became weakened and broke off. The fire burned all our 80-acre pasture in the SE¼ of section 27, the wood structures and grass on and around the King home site in the SW¼ of section 26, and some of the pasture and wooden structures including parts of the west door of the machine shed in the NE¼ of section 34. It also consumed the pasture in the NW¼ section 35, and the trees and remaining outbuildings on the Roy Crawford home site in that quarter, leaving behind only the charred remains of the trees and the tin roofs of a few sheds.

For decades the ubiquitous stone post and barbed wire fences the early settlers had worked so hard to install had symbolized this area of north-central Kansas, known as the "Land of the Post Rock." However, the gradual transition to increasingly larger combine headers and sprayer booms beginning in the late twentieth century prompted some farmers to remove no longer needed fences to provide more room for the larger equipment to maneuver. Also, unneeded fences served as breeding grounds for weeds that could then spread back into the fields. Delmar Hampl noted that by the early twenty-first century he had removed a total of seven miles of stone post fences for these reasons, and some of these fences had been on the Crawford land he and Kevin were farming.

The 2021 prairie fire damaged many of the remaining stone post and barbed wire fences, leading farmers to remove those damaged fences, and to replace them with new fences where needed. Federal disaster relief funding supported part of the costs of installing the new fences, but unfortunately, the new fences must be taller to meet current standards to receive FSA disaster support funding. Most of the stone posts are too short to meet those standards, so the new fences use steel or wooden posts that are not as picturesque as the stone posts that formerly characterized the area.

Piles of no-longer-wanted stone posts now dot the landscape across large parts of the Amherst area, serving as monuments to the early settlers who worked so hard and so skillfully to produce the miles of stone post fences that have now disappeared. These piles also remind us of the ever-present need to adapt to the changing conditions and new farming practices as farming on the Kansas prairie continues to evolve.

ACKNOWLEDGEMENTS

To tell this story, I have relied on family history, research into the times and communities which the various generations of my family witnessed, my own experiences, and some plausible assumptions to fill in the gaps. I made use of a variety of sources, including family documents, genealogies prepared by other family members, my own genealogical searches, family lore (written and oral), letters and notes, photographs, books, public land records, old newspapers, and internet articles. For those interested in pursuing aspects of this story in greater detail, I have included a set of notes and an extensive bibliography.

I was present during part of the period covered in this book, and those parts of this story are largely based on my memories and told from my perspective.

This book would not have been possible without the advice and encouragement I received from my wife, Charlotte Crawford, who patiently edited my numerous drafts, guiding me along my stumbling path to reach the final product. Thanks to her, the final product is far better than anything I could have produced on my own.

Once again, Post Rock Press skillfully designed and arranged the layout of this book, carefully including and positioning the numerous photographs at appropriate spots in the text. As before, the staff there performed this tedious work cheerfully, and with excellent results.

Several family members who preceded me showed the foresight to collect much of the Crawford, Hampl, Chegwidden, and Sechtem family genealogical, historical, and personal information that helped make this book possible. Special thanks are due in this regard to my aunt Pauline (Crawford) Bookstore, my mother Gladys (Chegwidden) Crawford, my dad Clarence Crawford, and my dad's cousin Delmar Hampl.

Thanks also to my sister Candace who has put up with me for these many years. She provided family lore, acumen, and details that helped stimulate my memory and contributed to fleshing out parts of this book.

Our family owes a huge debt to Delmar and Kevin Hampl for the excellent job first Delmar and later Kevin have done in farming our land and in continuing with the conservation and land improvement practices put in place while Dad was still actively farming. Both were also instrumental in providing information and insights that have found their way into this book. Each graciously agreed to read the manuscript, providing many helpful comments and a few factual corrections that I subsequently incorporated.

ACKNOWLEDGEMENTS

Newton King was a neighbor and friend while I was growing up on the farm, and remained a friend throughout his lifetime. I am indebted to him for freely sharing with me the results of some of his own genealogical and historical research into the Amherst community. Tom Beatty, another neighbor at whose house I spent many enjoyable evenings, was a long-term supporter of the Amherst and Luray communities and furnished helpful details about Luray history. Sadly, Newt and Tom are no longer with us, and both are now buried in the Amherst cemetery.

For more than ten years I have been fortunate to participate in genealogical and historical exchanges with John Oswald, my second cousin and a graduate of Luray High School. John supplied a wealth of information about the Sechtem family, the Hampl family, and the town and schools of Luray.

While researching the land transactions discussed in this book, I made numerous visits, as well as e-mail requests, to the Russell County Recorder of Deeds office. I thank the extremely helpful personnel there, who cheerfully guided me through the handwritten tomes that contained the information. Later, after digitization of more of their records, the process was simpler, and they quickly located and generated copies of the additional materials I wanted.

I also made numerous visits to the Russell County Historical Society over the years, where I was able to find highly informative notes and other historical information about Waldo and the Amherst community. I always found the staff there to be very helpful and courteous, and I thank them for the service they provided.

Whatever success I may have had in life, I owe at least in part to the teachers I had in the Luray schools, grades one through twelve. The school system was very small and had limited resources, but those teachers really cared and did their best to educate us to be responsible adults, even when our behavior made it appear that we were immune to education. They also deserve some of the credit for inspiring me to write this book.

NOTES

A New Beginning

Page

4-5 *The Scotch-Irish:* Dunaway, *The Scotch-Irish of Colonial Pennsylvania*, 1944; Northampton County Historical and Genealogical Society, *The Scotch-Irish of Northampton County, Pennsylvania*, 1926.

6 *French and Indian War:* Anderson, *Crucible of War: The Seven Years' War and the Fate of Empire in British North America, 1754–1766*, 2000.

7 *Northwest Territory:* Miller, Hunter, "British-American Diplomacy: The Paris Peace Treaty of September 30, 1783," accessed 2022; McCullough, *The Pioneers*, 2019; Indiana Historical Society, "Northwest Territory Papers and Documents," accessed 2022.

7-8 *Public Land Survey System:* White, C. Albert, *A History of the Rectangular Survey System*, 1983; Shipley, "list of US states' dates of admission to the union," 2020.

9 *Louisiana Purchase:* Lee, "The True Cost of the Louisiana Purchase," 2017; Point to Point Surveyors, "The Louisiana Purchase Survey," 2016.

11 *Kansas Territory and Statehood:* Kansas Historical Society, "Kansas Territory," 2010; Alexander, "Territorial Kansas & the Struggle for Statehood," 2022.

11-13   *Land Survey in Kansas:* Miner, *West of Wichita*, 1986; Suchy, "The Public Land Survey System in Kansas," 2002.

13-14   *Homestead Acts:* Archives.gov, "Homestead Act of 1862," accessed 2022.

14-15   *Railroads:* Weiser, "Kansas Pacific Railway," 2020; Library of Congress, "Map of Nebraska Showing the Union Pacific Railroad Land Grant," 1880.

15   *Population data:* United States Census Bureau, "US Census Data." All population values specified in this narrative are based on US Census data unless otherwise noted.

15-16   *Kansas climate and weather:* National Weather Service, "Origin of Wind," accessed 2022; National Weather Service, "Hemispheric Prevailing Winds," accessed 2022; Conrad, "How Do the Rocky Mountains Influence Climate," 2020.

First Generation

21-22   *Osborne County History:* Osborne County, "The Creation and Organization of Osborne County Kansas," accessed 2021; Cutler, *History of the State of Kansas*, Osborne County, Part 2, 1883; Ise, *Sod and Stubble*, 1996. *Sod and Stubble* presents a moving account of what life was like for the Ise family, who settled in Osborne County in 1873.

22   *Train fare:* Trimble, "What was the Fare for Railroads and Stagecoaches?" 2018.

22   *Wages:* Bureau of Labor and Industrial Statistics, *Fourth Annual Report*, 1889.

22-23   *Russell and Russell County:* Cutler, *History of the State of Kansas*, Russell County, Part 1; Part 2; Part 3; Part 5, 1883.

24   *Saline River:* Root, "Ferries in Kansas, Part VII, Saline River," 1935.

24-25   *Prairie grasses and plants:* Wrangle, "North American Mixed Grass Prairie," 2022; Hampl, Delmar & Kevin Hampl, Discussion October 28, 2022.

25   *Travel to Russell:* The Crawford party might have equipped themselves in the Kansas City area and then driven their

wagons full of equipment to Russell. However, there is some evidence that at least the Joseph King family first came to Russell by train, and since they settled in the Amherst area before the Crawfords did, it is reasonable to assume that the Crawfords followed this somewhat easier route as well. King, E-mail to Kent Crawford July 28, 2012.

25 *Price of horses:* Horses, "How Much Did a Horse Cost in the 1800s?" Accessed 2022.

25 *Prices of farm machinery:* University of Missouri Libraries, "Price and Wages by Decade," accessed 2021.

28-29 *Figure 4 map locations and dates:* NorthWest Publishing Co., *Plat Book of Russell County Kansas*, 1901. Locations are based on the *Plat Book*; dates are based on land records or Crawford family knowledge. The date given for the Roy Crawfords is the date they moved into their new house on section 35. The date given for the Alex Hampls is the date they moved into their house in section 36.

30-31 *Water:* Hampl, Delmar & Kevin Hampl, Discussion October 28, 2022; Hampl, Delmar, Notes to Kent Crawford, November 25-29, 2010.

31-33 *Building a sod dugout house:* Bookstore & Crawford, Discussion May 2008; Soucy, "Dugouts and Sod Houses— How to Build Them," 2010; American History, "Life in a Sod House," accessed 2022.

32 *Moldboard plow:* Your Local Farmer, "The Moldboard Plow: history, uses, and definitions," 2020.

34-36 *Farming operations:* Bellis, "History of American Agriculture," 2019; Brainard, "History of Kansas Agriculture," accessed 2021; Living History Farms, "Grain Harvest and Threshing Time," 2015.

36 *Garden:* Ise, *Sod and Stubble*, 1996.

37-38 *Quarrying stone:* We still have a set of Crawford family feathers and wedges, possibly the ones that Harmon originally used.

38-39 *Stone house:* Bluestem Quarry and Stoneworks, "Kansas Limestone History," 2007; Bookstore & Crawford, Discussion May 2008; Luder, Laura, Letters to Pauline (Crawford) Bookstore, 1969. Laura (Crawford) Luder was not yet born at the time this house was built, thus some of this is her recollections of what her parents had told her. However, she did live in that house for many years, so most of the details she provided were based on her first-hand knowledge.

39 *Livestock shelters:* Luder, John, "The Golden Prairies of the Middle West," no date.

41 *Figure 7:* This figure was derived from a 1919 photograph of the house, modified to conform to the earlier configuration recounted by Laura (Crawford) Luder.

41-42 *Thomas Beatty:* Luder, Laura, Letters to Pauline (Crawford) Bookstore, 1969. Laura was not yet born at this time so this anecdote was based on her recollections of what her parents had told her.

43 *Death of Joseph Franklin Crawford:* Bookstore, Crawford Genealogy Notebook, clipping from unidentified newspaper January 6, 1881.

43 *Rattlesnakes:* Luder, Laura, Letters to Pauline (Crawford) Bookstore, 1969. Laura was not yet born at this time so this anecdote was based on her recollections of what her parents had told her.

43-44 *Whiskey:* Bookstore, Discussion October 2010. McHenry whiskey was distilled and bottled in Columbia County Pennsylvania.

44 *Candus:* Luder, Laura, Letters to Pauline (Crawford) Bookstore, 1969. Laura was not yet born at this time so this anecdote was based on her recollections of what her parents had told her.

44 *Jacob Bean:* Luder, Laura, Letters to Pauline (Crawford) Bookstore, 1969. Laura was not yet born at this time so this

anecdote was based on her recollections of what her parents had told her.

45   *Dust storms:* Malin, "Dust Storms: Part Two, 1861-1880," 1946.

45-46   *Blizzard of 1885:* Schaller, *et al.,* "Climate effects of the 1883 Krakatoa eruption: Historical and present perspectives," 2009; Giles, "Remembering the frigid winter of 1885," 2016; Holt, "Victorian Blizzards, Nonstop in the 1880s," 2017.

46   *Catharine O'Connor:* Luder, Laura, Letters to Pauline (Crawford) Bookstore, 1969. Laura was not yet born at this time so this anecdote was based on her recollections of what her parents had told her.; Find-A-Grave, "Daniel James O'Connor (1843-1915)," accessed 2021; United States Census Bureau, "US Census Data."

46-47   *Blizzard of 1886:* Kansapedia, "Blizzard of 1886," 2011.

47   *Winter Clothing:* Robinson, "Winter Survival Skills That Kept The Pioneers Alive," accessed 2022.

48   *Crawford farm inventory (1885):* Kansas, "1885 Kansas Agricultural Census," 1885; US Department of Agriculture, *Kansas Wheat History,* 2017.

49-50   *Post rock fence:* Bluestem Quarry and Stoneworks, "Kansas Limestone History," 2007; Crawford, Clarence, "Building a Post Rock Fence," 2021; Hampl. Delmar, Notes to Kent Crawford, October 28, 2022; Hampl, Delmar & Kevin Hampl, Discussion October 28, 2022.

50-51   *Salina, Lincoln, and Western Railway Company:* Ridgway, "Remembering Union Pacific's Plainville Branch," 1998; Nothern, "Cattle Trails, Rodeos & Livestock Stories," accessed 2022; Abandoned Rails, "The Plainville Branch," accessed 2022; Trainorders, "UP: Remembering the Plainville Branch," 2019.

50-51   *Luray: Russell Daily News,* "J. W. Van Scoyoc Third to File Claim in Russell County," 1961.

51   *Waldo:* Tripp, "Do You Remember when Waldo Had the Following?" 1970; Clow, "City of Waldo, Kansas," 2003.

52-54   *Crawford Homestead Patent:* Russell County Recorder of Deeds office, Land records; Crawford, Harmon, Homestead Patent Document, 1890; Ise, *Sod and Stubble*, p. 108, 1996.

53-54   *Retreat to Kansas City:* Luder, Laura, Letters to Pauline (Crawford) Bookstore, 1969. Laura was not yet born at this time so this anecdote was based on her recollections of what her parents had told her; Bookstore, Crawford Genealogy Notebook, newspaper clipping of Ruth Candus (Dildine) Crawford Norris obituary.

54   *Quindaro:* Legends of Kansas, "Quindaro, Kansas–a Free-State Black Town," accessed 2022. Quindaro was a port town founded by free-staters on the Kansas bank of the Missouri river. It was a booming town from 1856 to 1861, but it underwent a rapid decline following the start of the Civil War and its town charter was removed in 1862. After the completion of the Civil War the abandoned townsite was occupied by former slaves seeking to escape the Jim Crow South.

54   *Quindaro School:* KCKPL, "Historic School Buildings of Kansas City, Kansas," accessed 2022. Quindaro continued to be a largely black community for several years, but since schools were segregated at that time, the area had a separate "colored'" one-room elementary school in addition to the all-white one-room Quindaro Elementary School. The Quindaro area was eventually incorporated as part of Kansas City, Kansas.

55   *Return to the homestead:* Luder, Laura, Letters to Pauline (Crawford) Bookstore, 1969.

55-56   *Droughts in the 1890s:* Blackmar, *Kansas, a Cyclopedia of State History*, Volume I, pp. 547-549, 1912.

55-57   *Land ownership changes:* Russell County Recorder of Deeds office, Land records.

57   *Crawford farm inventory (1895):* Kansas, "1895 Kansas Agricultural Census," 1895.

57-58   *Dust storms:* Malin, "Dust Storms: Part Three, 1881-1900," 1946; Luder, Laura, Letters to Pauline (Crawford) Bookstore, 1969.

58   *Blizzard of 1899:* Vogan, "A LOOK BACK Greatest of Arctic Outbreaks: 1899," 2015; Miller, Kaitlin, "The Most Devastating Winter Storms in US History," 2020; Luder, Laura, Letters to Pauline (Crawford) Bookstore, 1969.

59   *Hunting:* Luder, Laura, Letters to Pauline (Crawford) Bookstore, 1969.

59   *Crawford orchard:* Luder, Laura, Letters to Pauline (Crawford) Bookstore, 1969; Kansas, "1895 Kansas Agricultural Census," 1895.

59-60   *Entertainment:* Luder, Laura, Letters to Pauline (Crawford) Bookstore, 1969.

60-62   *Amherst School and Amherst Church:* Batt, *The History of the Russell County Rural Schools*, 1994; Bookstore, Crawford Genealogy Notebook, "Brief History of Amherst," accessed 2008; Russell County Historical Society, Notes about Waldo history.

62-64   *Farm implements:* University of Missouri Libraries, "Price and Wages by Decade," accessed 2021; Miller, Lynn, "A Short History of the Horse-Drawn Mower," accessed 2022; Bellis, "History of American Agriculture," 2019; US Department of Agriculture, "Grain Drills and Planters," 2016; Wendel, *Encyclopedia of American Farm Implements and Antiques, 186,* 2004.

64-66   *Threshing machine:* EDinformatics, "Threshing Machine," accessed 2022; Hansen, "How a Threshing Machine Works," accessed 2021; University of Missouri Libraries, "Price and Wages by Decade," accessed 2021; McManus, "Steam Threshing Rings," accessed 2021; Penner, *Section 27,* 2002.

65   *Wheat acreage:* US Department of Agriculture, *Kansas Wheat History,* 2017.

66   *Crawford Land purchases:* Russell County Recorder of Deeds office, Land records.

67-68 *Luray:* NorthWest Publishing Co., *Plat Book of Russell County Kansas*, 2001; Kansapedia, "Opera Houses," accessed 2021; *Luray Herald*, "Moving Picture Show," 1902; Encyclopaedia Britannica, "Kinetoscope," 2020; Oswald, Communication November 2015, 2020. The opera house picture came from Eldon Hampl, who lived in an apartment in this building in later years, and operated an electrical appliance sales and repair shop there. This picture was provided by Eldon's nephew, John Oswald, who also provided useful comments about it.

69 *Waldo:* Tripp, "Do You Remember when Waldo Had the Following?" 1970; Russell County Historical Society, Notes about Waldo history, no date; Blackmar, *Kansas, a Cyclopedia of State History*, Volume II, p. 859, 1912; NorthWest Publishing Co., *Plat Book of Russell County Kansas*, 1901.

69 *Roads:* Woods, *County Road Laws of Kansas*, 2008; NorthWest Publishing Co. *Plat Book of Russell County Kansas*, 1901. The state laws went on to specify how much right-of way counties could extract from the adjacent land owners, and stated that such county roads must be designed to extend equal amounts on each side of the sectional line. The acts also set forth a process by which the residents could petition the county commissioners to open such a road. The first of these acts was passed in 1856, but it was not until 1873 that the state legislature passed the first such act to specifically include Russell County.

69-70 *Harmon Crawford family:* Bookstore, Crawford Genealogy Notebook.

71 *Roy Crawford land:* Bookstore & Crawford, Discussion May 3, 2008; Russell County Recorder of Deeds office, Land records.

71 *John Pospishil:* Bookstore & Hampl, Hampl Genealogy Notebook. John Pospishil was a member of the Pospishil family that emigrated from Bohemia to West point, Nebraska. Joseph Pospishil, one of John's brothers, married

Christina Hampl, one of Alex Hampl's sisters. Frank Pospishil, another brother of John, settled near Luray, and may have been involved in naming the Pospishil Opera House there.

71   *Figure 14 quarry:* Hampl, Delmar, Notes to Kent Crawford, October 28, 2022.

Second Generation

77-81   Hampl family: Bookstore & Hampl, Hampl Genealogy Notebook; Hampl & Hubbard, Oral interview 1980.

77   *Emigration from Bohemia:* Kysilka, "Emigration to the USA from the Policka region in 1850—1890," 1999; My Czech Roots, "Emigration—Basic History," accessed 2021; Czech Friends, "Why they went—Reasons for Czech emigration to America in the 19th century," accessed 2021; Encyclopaedia Britannica, "Czechoslovakia," 2020.

78   *Sandburs:* Hampl, Delmar & Kevin Hampl, Discussion October 28, 2022. Sandburs were not common in eastern Nebraska where the Hampls had been living.

81   *Harmon Crawford death:* Bookstore, Crawford Genealogy Notebook; Russell County Recorder of Deeds office, Land records.

81-82   *Marriage of Roy Crawford and Albina Hampl:* Bookstore, Crawford Genealogy Notebook.

83-85   *Laura, Charlie and Candus:* Bookstore, Crawford Genealogy Notebook; Luder, Laura, Letters to Pauline (Crawford) Bookstore, 1969.

85-86   *Hampl family:* Bookstore & Hampl, Hampl Genealogy Notebook; Hampl, Delmar, Notes to Kent Crawford November 25-29, 2010.

86   *Death of Thomas Beatty:* Find-A-Grave, "Thomas Anderson Beatty (1830-1909)," accessed 2021; Russell County Recorder of Deeds office, Land records.

87   *Roy, Albina, and family:* Bookstore, Crawford Genealogy Notebook.

87-91   *Roy Crawford farm:* Luder, Laura, Letters to Pauline (Crawford) Bookstore, 1969; Russell County Recorder of Deeds office, Land records; Bookstore, Discussion July 31, 2011; Bookstore, Telephone discussion February 21, 2010.

91-92   *Forge:* Hampl. Delmar, Notes to Kent Crawford, October 28, 2022.

93   *Travel:* Hampl & Hubbard, Oral interview 1980; Bookstore, Telephone discussion February 21, 2010.

93-94   *Winter in Rocky Ford:* Bookstore, Discussion July 31, 2010; Waymarking, "History of Rocky Ford," 2011; US Department of Agriculture, *Kansas Wheat History,* 2017.

94   *Roy Crawford automobile:* Classic & Collector Cars, "1919 Ford Model T Touring," accessed 2022; Bookstore, Handwritten notes April 7, 2009.

96   *Car races:* Hampl & Hubbard, Oral interview 1980.

96-98   *Crawford trip west:* Crawford, R. Kent, *Ruts, Guts, & a Model T Truck: Cruising the West at 15 Miles per Hour,* 2021.

98-99   *Summer fallow:* Hampl, Delmar, Discussion July 30, 2021.

100   *Chudomelkas in Figure 27:* Bookstore & Hampl, Hampl Genealogy Notebook. One of the sisters of Alex Hampl Sr. had married a Chudomelka, and their families kept in close touch even though the Chudomelkas still lived in Nebraska. However, the Chudomelkas in this picture were not identified.

103   *Fishing:* Bookstore, Discussions July 30-31, 2010.

104-106   *Luray and Waldo:* Blackmar, *Kansas, a Cyclopedia of State History,* Volume II, p. 193, 1912; Tripp, "Do You Remember when Waldo Had the Following?" 1970; Connelley, *A Standard History of Kansas and Kansans,* 1919.

106-107   *Pauline and Clarence:* Bookstore, Telephone discussion February 21, 2010; Bookstore, Discussions July 31, 2011.

108-109   *Carpooling to High School:* Bookstore, Discussion July 30-31, 2010. Gladys and Eulah Bratton were daughters of Albert Bratton, who lived about a mile and east of the Roy

Crawfords' house, and the Russell Cochrun family lived near the Brattons.

108-111   *College:* K-State Alumni Association, "K-State History," accessed 2021; Pearson & Atucha, "Agricultural Experiment Stations and Branch Stations in the United States," 2015; Kansas State University, "What is K-State Research and Extension?" Accessed 2021; Bookstore, Crawford Genealogy Notebook, clipping from unidentified newspaper "Luray Girl Chosen Ag Princess," 1934; US Department of Agriculture, "The Home Demonstration Agent," 1951; Kansas State University, "K-State Army ROTC History," 2018; Kansas State College, *The Royal Purple for 1936*; Bookstore, Discussion July 30-31, 2010.

113   *Binder:* Reinhardt, "Harvesting Wheat," accessed 2021; University of Missouri Libraries, "Price and Wages by Decade," accessed 2021.

114-115   *Header:* Isern, "Adoption of the Combine on the Northern Plains," 1980.

115-118   *Tractors and threshing:* White, William J., "Economic History of Tractors in the United States," 2008; Isern, "Adoption of the Combine on the Northern Plains," 1980; Tractor Data, "Aultman & Taylor 30-60," accessed 2021; Revivaler, "Aultman and Taylor 30-60," accessed 2021; Yesterday's Tractor Co., "Antique Tractor Resources—Original Tractor Prices: Case," accessed 2021; Bookstore, Discussion July 31, 2011.

118-119   *Roy's tractor:* Fandom, "McCormick Deering 15-30," accessed 2021.

119-120   *Combine:* Herring, "Harvest Equipment: A Brief History of the Combine," 2020; The History Museum, "The Oliver Corporation," accessed 2021. The Nichols & Shepard Company began making threshers around 1900. Somewhat later, they designed and created this line of horse-pulled or tractor-pulled combines. The Nichols & Shepard Company merged with the Oliver Corporation in 1929. For several years after the merger of 1929, the Oliver Company continued to manufacture an entire line of combines first developed by Nichols and Shepard.

120-121 *One-way:* Kuhn North America, "History in North America," accessed 2021; South Dakota State University, "One-Way Disc Plow—Adjustment and Operation," 1977; University of Minnesota, "The One-Way Disc Tiller," 1938; Crawford, Roy, Roy Crawford estate tax forms, personal property list, 1960.

122-125 *Economics of farming:* US Department of Agriculture, *Kansas Wheat History*, 2017; Russell County Recorder of Deeds office, Land records; United States Census Bureau, "US Census Data."

125-126 *The Great Depression:* Find-A-Grave, "John A. O'Leary, Sr. (1901 - 1984)," accessed 2022; Schultz family, *Luray High School 1914-1977, Alumni Memory Book*, 2012; US Department of Agriculture, *Kansas Wheat History*, 2017; United States Census Bureau, "US Census Data."

126-128 *Dust Bowl:* Ganzel, "The Dust Bowl," 2003; Bookstore, Discussion, date not recorded; Hampl, Delmar, Notes to Kent Crawford, November 25-29, 2010; Hampl, Sarah, Sarah (Griffin) Hampl notes about her life, date unknown.

128-129 *Jackrabbit drives:* Kansas Historical Society, "Jackrabbit Drives," 2015.

129 *Crawford land:* Russell County Recorder of Deeds office, Land records.

131-132 *Crawford and Hampl families:* Bookstore, Crawford Genealogy Notebook; Bookstore & Hampl, Hampl Genealogy Notebook.

132-137 *New Deal:* Rasmussen, *et al.*, *A Short History of Agricultural Adjustment*, 1976; The Living New Deal, "Agricultural Adjustment Act (1933, reauthorized 1938)," 2016; Lotterman, "Farm Bills and Farmers: The Effects of Subsidies Over Time," 1996; National Agricultural Law Center, "Conservation Programs—An Overview," accessed

2021; History Colorado, "WPA Privy (1935-1943)," accessed 2021; Hampl, Delmar, Notes to Kent Crawford, November 25-29, 2010.

137-140 *Economics of Farming:* Kansas Department of Agriculture, "Kansas Farm Facts," 2019; Encyclopedia.com, "Farm Foreclosures," accessed 2021; US Department of Agriculture, *Kansas Wheat History*, 2017; Farmland Information Center, "Kansas Data and Statistics," accessed 2021; K-State Research and Extension, *Wheat Production Handbook*, 1997; Dalrymple, "Changes in Wheat Varieties and Yields in the United States, 1919-1984," 1988; Martin, "A Short History of Prices, Inflation since the Founding of the US,"; 2017. McMahon, "Historical Consumer Price Index (CPIU) Data," accessed 2021; Webster, "Inflation Calculator," 2021.

Third Generation

145-147 *John Chegwidden and Mary (Lee) Chegwidden:* Crawford, Gladys, Chegwidden Genealogy Notebook; Crawford, Gladys, Gladys (Chegwidden) Crawford notes about her life, 1993; Russell County Recorder of Deeds office, Land records; *Russell Record*, "House of Seven Gables," 1977.

147-148 *Caspar Sechtem and Mary (Essig) Sechtem:* Crawford, Gladys, Sechtem Genealogy Notebook; Crawford, Gladys, Gladys (Chegwidden) Crawford notes about her life, 1993; Military History Matters, "The Franco-Prussian War," 2020; Sechtem, Caspar Sechtem army discharge papers, 1877; Russell County Recorder of Deeds office, Land records.

148-149 *William Chegwidden and Sophia (Sechtem) Chegwidden:* Crawford, Gladys, Chegwidden Genealogy Notebook; Crawford, Gladys, Sechtem Genealogy Notebook; Russell County Recorder of Deeds office, Land records; Dorrance and Lucas are both small towns in the eastern part of Russell County.

150 *Gladys and Helen:* Crawford, Gladys, Gladys (Chegwidden) Crawford notes about her life, 1993.

151-152 *Clarence Crawford and Gladys (Chegwidden) Crawford:* Bookstore, Crawford Genealogy Notebook; Crawford, Gladys, Gladys (Chegwidden) Crawford notes about her life, 1993.

153-156 *World War II:* Crawford, Gladys, Gladys (Chegwidden) Crawford letters to her sister Helen Hampl, 1941; Crawford, Clarence, Clarence Crawford letters to Gladys Crawford, 1941-1944; ozatwar.com, "94th Coast Artillery (AA) Regiment 40th Anti-Aircraft Brigade in Australia During WW2," accessed 2014; Kauffman, "209th AAA Automatic Weapons Battalion (Self-Propelled)," 2001; World War II Troopships, "Queen Mary—Specific Crossing Information—1942," 2007.

154 *Roy Crawford land:* Russell County Recorder of Deeds office, Land records.

157 *Pauline marriage: St. John News,* "Bookstore-Crawford Wedding," 1945.

163-164 *Telephone:* The Museum of Yesterday, "Antique Wireless Equipment Collection—Telephone," accessed 2022.

165-166 *Rural Electrification Act:* Rural Electrification Administration, *Rural Lines, USA: the story of the Rural Electrification Administration's first 25 years, 1935-1960,* 1960; Rural Electrification Administration, "Rural Lines—USA, The Story of Cooperative Rural Electrification," 1981.

167-168 *Conservation plan (1948):* Soil Conservation Service, "Conservation Plan, Clarence R. Crawford Operator, Roy A. Crawford Owner, 1948." This plan was quite detailed, but the main points were: (1) Legumes will be used in the rotation for cropland; (2) All crop residue, stalks and stubble will be incorporated at or near the surface of the soil to increase organic content and to protect the land from wind and water erosion; (3) All tillage and seeding operations on the sloping lands will be on the contour, parallel to terraces;

(4) Waterways will be prepared approximately at the locations shown on the land use map; (5) The farmer agrees to construct gradient terraces emptying onto grassed waterways on the cropland. (6) One stock water pond will be constructed in the pasture on section 32 at the location shown on the land use map.

167-168   *Terraces:* US Department of Agriculture, *Engineering Field Handbook*, "Chapter 8, Terraces," 2011; Powell & McVay, *Terrace Maintenance*, 2004; Hampl. Delmar, Notes to Kent Crawford, October 28, 2022.

168-169   *Experiment Stations:* Zhang, "Wheat Breeding," accessed 2021; US Department of Agriculture, *Kansas Wheat History*, 2017; Reitz & Laude, "Comanche and Pawnee: New Varieties of Hard Red Winter Wheat for Kansas," 1943.

169-170   *Luray School buildings:* Batt, *The History of the Russell County Rural Schools*, 1994; Schultz family, *Luray High School 1914-1977, Alumni Memory Book*, 2012.

176   *Custer's Last Stand:* National Museum of American History. "Chromatolithograph entitled 'Custer's Last Fight,'" accessed 2023.

177   *Wheat statistics:* US Department of Agriculture, *Kansas Wheat History*, 2017.

178-179   *Custom cutters:* Lutz & Aschwege, "Custom Harvesting Rates Paid in Nebraska 1958," 1959.

179   *Death of William Chegwidden:* Crawford, Gladys, Chegwidden Genealogy Notebook.

179-180   *Truck and combine:* Auto editors of Consumer Guide, "1948-1952 Ford F-Series Trucks," accessed 2021; Thompson, "My Massey combines from 1948 to 1991," 2020; Hampl, Delmar & Kevin Hampl, Discussion October 28, 2022.

181   *Land transfers:* Crawford, Roy, Roy Crawford gift tax documents, 1950; Russell County Recorder of Deeds office, Land records. This land was the part of the farm that Grandpa Crawford intended for Dad to inherit. However, these land transfers were made prior to Grandpa's death

and were treated as a gift, so appropriate gift taxes were paid.

182 *Case tractor:* Yesterday's Tractor Co., "Antique Tractor Resources—Original Tractor Prices: Case," accessed 2021; Bookstore & Hampl, Hampl Genealogy Notebook, "Harvey E. Bean Obituary," 1955. Leonard Bean was one of Dad's cousins, the son of Harvey Bean and Mary (Hampl) Bean, Albina's sister.

182-183 *Pond south of our house:* Hampl, Delmar, Notes to Kent Crawford, October 28, 2022. By the 1980s, silt had filled this pond, so Delmar removed the dam to return the area to grass and hay. He used some of the dirt from the dam to widen many of the field entrances to accommodate larger equipment and trucks.

187-188 *Blade plow or undercutter:* Kantor, "Conserving Soil and Water in Dryland Wheat Region," 2015; Steel in the Field, "Stubble Mulch Blade Plow," 2001.

189-191 *Fertilizer:* Hergert, *et al.,* "A Historical Overview of Fertilizer Use," parts 1, 2 and 3, 2015; Ganzel & Reinhardt, "Postwar Fertilizer Explodes," accessed 2021; Hampl, Delmar & Kevin Hampl, Discussion October 28, 2022.

191-192 *Farm programs:* Rasmussen, *et al., A Short History of Agricultural Adjustment,* 1976.

192-193 *Flood:* Juracek, *et al.,* "The 1951 Floods in Kansas Revisited," 2001; Legends of Kansas, "Black Friday Flood, 1951," accessed 2022; US Department of Agriculture, *Kansas Wheat History,* 2017.

193-194 *Drought:* Nace & Pluhowski, "Drought of the 1950's with Special Reference to the Midcontinent," 1965; National Drought Mitigation Center, "The Dust Bowl," accessed 2021.

194 *Blizzard:* National Weather Service, "60th Anniversary of the March 23-25, 1957 Blizzard," 2021.

195 *Milo:* Carter, *et al., Grain Sorghum (Milo), Alternative Field Crops Manual.* 1989; Kansas State University, *Grain Sorghum Production* Handbook, 1998.

196 *Luray bank:* Find-A-Grave, "John A. O'Leary Sr. (1901-1984)," accessed 2022; Find-A-Grave, "John A. O'Leary Jr. (1926-1989)," accessed 2022.

196-197 *Slip-form elevator and watertower:* Cart, "Johnson-Sampson Construction Company," accessed 2022; Baxter, "The Slip Form Method and Reinforced Concrete Grain Elevators," accessed 2022; Schmitt, "The Shape of Water Towers—An Engineering History," accessed 2022. The Luray elevator and water tower were probably constructed by the Johnson-Sampson Construction Company of Salina, Kansas.

Fourth Generation

204-205 *Death of Roy Crawford:* Bookstore, Crawford Genealogy Notebook; Crawford, Roy, Roy Crawford will.

205-206 *Roy Crawford house and barn:* Hampl, Delmar, Notes to Kent Crawford, November 25-29, 2010; Hampl, Delmar & Kevin Hampl, Discussion October 28, 2022. The Crawfords' cistern had been huge, and after the house was removed, Delmar Hampl had the cistern filled with 30 yards of sand to prevent it from becoming a safety hazard.

206-207 *Clarence Crawford land transfers:* Russell County Recorder of Deeds office, Land records.

207 *The move to Russell:* Crawford, Gladys, Gladys (Chegwidden) Crawford notes about her life, 1993.

207 *Death of Albina (Hampl) Crawford:* Bookstore, Crawford Genealogy Notebook.

208 Soil and Water Conservation Plan (1965): Soil Conservation Service, "Soil and Water Conservation Plan, Clarence R. Crawford Operator, Pauline Bookstore Owner, 1965." The plan specified that: "(1) Farming operations will be on the contour as soon as terraces are constructed; (2) Fallow, small grains and sorghums with a legume and/or commercial

fertilizer will be used to increase fertility and organic matter; (3) Crop residue will be left on or near the soils surface to help control wind and water erosion; (4) Grassed Waterways to be shaped and seeded to a Brome-Western Wheat grass mixture in a sorghum cover; (5) Balance of terraces will be constructed after waterways are well established; and (6) Hayland will be returned to native grass for proper land use. Native grass will be managed to maintain and improve the present composition of native grass."

208-209 *Life in Russell:* Crawford, Gladys, Gladys (Chegwidden) Crawford notes about her life, 1993.

208 *Teaching school in Russell:* Crawford, Gladys, Gladys (Chegwidden) Crawford notes about her life, 1993; Gould, "Cemetery Folklore Studied," 1981.

209 *Russell County Extension Council: Russell Daily News*, "Luray Man New Member of State Education Board," 1959; K-State Research and Extension, "Chisholm Trail District," accessed 2022.

209 *State Board of Education: Russell Daily News*, "Luray Man New Member of State Education Board," 1959. At that time the Kansas State Board of Education was a seven-member non-partisan board including one member from each of the six Congressional Districts and one from the state at large. The governor appointed members for terms of three years, with a maximum of two such terms. Their duties included approving or rejecting school textbooks, the issue and renewal of certificates of teachers, curricula, and courses of study, and acting in an advisory capacity to the state superintendent of public instruction.

209-210 *Candace—school, career, and marriage:* Bookstore, Crawford Genealogy Notebook, clipping from unknown newspaper, "Crawford-Keiderling Wedding on Sept 4," 1976.

211 *Goodyear Award: Russell Daily News*, "Goodyear Award presented to Clarence Crawford," 1976; Agronomy

eUpdates, "Kansas Bankers Association Conservation Award Program," 2020.

211-212 *Soil and Water Conservation Plan (1988):* Soil Conservation Service, "Soil and Water Conservation Plan, Clarence R. Crawford Trust #1 Operator, Pauline Bookstore Owner, 1988." The plan stated that: (1) A cropping system of sorghum-fallow-wheat or its equivalent will be followed on these fields; (2) A minimum of 30% residue cover from the previous crop will remain on the soil surface at planting time; (3) All tillage and planting operations are performed on the contour using the terraces as guidelines; (4) Tillage is managed to leave sufficient residue on the soil surface to control soil loss from wind erosion. During the critical wind erosion period, a minimum of 250 lb. small grain equivalent of wheat residue or 500 lb. small grain equivalent of sorghum residue will be maintained on the soil surface. After the critical wind erosion period has passed, residue is gradually incorporated into the soil leaving no less than 10% residue on the soil surface at planting time; (5) Waterways to be seeded, fertilized (if needed); (6) Terraces are to be maintained at or above 1.0-foot height between channel and ridge.

212 *Figure 63:* Tim Keiderling is married to Candace, who kept her maiden name. Michael Keiderling is their son (age 5 at the time of this photo). Lara and Clare are Charlotte's and my daughters (ages 19 and 16 at the time of this photo).

213 *Land transfer to Kent and Candace Crawford:* Russell County Recorder of Deeds office, Land records.

213 *Pauline recognized for years of community service:* Bookstore, Crawford Genealogy Notebook, clipping from unidentified newspaper, "Pauline Bookstore named Xi Zeta Eta 'Lady of the Year,'" 2000.

213-215 *Farming boom and bust:* Rasmussen, *et al., A Short History of Agricultural Adjustment, 1933-75,* 1976; Lotterman, "Farm Bills and Farmers: The effects of subsidies over time," 1996;

US Department of Agriculture, *Kansas Wheat History*, 2017; Ganzel, "Farm Bust of the 1980s." 2009.

216 *Deaths of Bertha Maude and Lloyd Beatty:* Find-A-Grave, "Bertha Maude Reiss Beatty (1914-1982)," accessed 2022; Find-A-Grave, "Lloyd Raymond Beatty (1912-1984)," accessed 2022.

216 *Loss of the railroad:* Trainorders.com, "UP: Remembering the Plainville Branch," 2019; Abandoned Rails, "The Plainville Branch—Salina to Oakley, KS," accessed 2022

216-219 *Loss of population:* United States Census Bureau, "US Census Data," Schultz family, *Luray High School 1914-1977, Alumni Memory Book*, 2012.

217-218 *School Consolidations:* Kansas High School Football History, "Lucas-Luray Cougars," accessed 2022; *Topeka Capital Journal*, "Schools ask to consolidate," 2010; Schultz family, *Luray High School 1914-1977, Alumni Memory Book*, 2012; *Plainville Times*, "Natoma USD-391 & Paradise-Waldo USD-399 School Patrons Favor Joining Districts," 1974; Clow, "City of Waldo, Kansas," 2011.

217 *Football championship:* Schultz family, *Luray High School 1914-1977, Alumni Memory Book*, 2012.

218 *Changes at the bank:* Find-A-Grave, "John A. O'Leary Jr. (1926-1989)," accessed 2022; Bennett, "Saving a Town," 2001; Bankencyclopedia.com, "The Peoples State Bank of Luray, Kansas in Luray, Kansas (KS)," accessed 2022; US Bank Locations, "The Peoples State Bank of Luray, Kansas," accessed 2022; UMB Bank, letter to Roy Kent Crawford, June 20, 2011; US Bank Locations, "The Farmers Bank of Osborne," accessed 2022.

219-220 *Century Farm designation:* Kansas Farm Bureau, "Do You Live on a Century Farm?" Accessed 2021; Crawford, Clarence, Century Farm Citation, 2001.

227 *Luray Friendship Day and Luray Fish Fry:* Russell County Economic Development & Convention and Visitors Bureau, "56th Annual Luray Friendship Day," accessed 2022; Blue

Hills Heritage Foundation, "—2018 Event Inductee—Luray Fish Fry," accessed 2021.

233   *Farm machinery costs:* Thompson, "My Massey combines from 1948 to 1991," 2020; Ganzel, "Harvest Technology," 2009; US Department of Agriculture, *Kansas Wheat History*, 2017.

233-235   *1948-2010 comparison:* Hampl, Delmar & Kevin Hampl, Discussion October 28, 2022; Other data based on figs. 48 and 49. Different choices for the profit margins or for the wheat prices or yields would of course change the absolute number of bushels or acres required, but the gist of the argument remains the same.

235   *Chemical costs:* US Department of Agriculture, "Fertilizer Use and Price," 2019; US Department of Agriculture, "Commodity Costs and Resources," 2021; US Department of Agriculture, "ERS Charts of Note," fertilizers and pesticides, 2021.

235-236   *Effects of a changing climate:* US Global Change Research Program, *Fourth National Climate Assessment*, 2018; Union of Concerned Scientists, "Climate Change and Agriculture—A Perfect Storm in Farm Country," 2019; United States Environmental Protection Agency, "What Climate Change Means for Kansas," 2016.

236-237   *Automation:* Mottech, "Why Automating Agriculture is the Future of Farming," 2022; Bedford, "How Automation Will Transform Farming," 2017; IDTechEX.com, "Agricultural Robots and Drones 2017-2027: Technologies, Markets, Players," 2017.

237-238   *Alternative crops:* Dreibus, "Goodbye, Kansas Wheat?" 2018; Bounds, "Kansas: A Leader in Wheat, Grain Sorghum, and Beef Production," 2021; McFadden, "Drought-Tolerant Corn in the United States: Research, Commercialization, and Related Crop Production Practices," 2019. In 2017 Kansas farms produced 319 million bushels of wheat, 194 million bushels of grain sorghum, 694 million bushels of corn, 197 million bushels of soybeans, 74 million pounds of sunflowers, 5.6 million tons of forage, and even 188 thousand bales of cotton.

238   *Hard White Winter Wheat:* Paulsen, "Hard White Winter Wheat for Kansas," 1998; Kansas Ag Growth, "Wheat," accessed 2022.

238-239   *Perennial wheat:* The Land Institute, "Kernza Grain," 2022; Snapp & Marone, "Perennial Wheat," 2014.

239   *Cropland into grassland:* US Department of Agriculture, "Conservation Reserve Program," 2022; Nickel, "Converting Marginal Cropland to Perennials Builds Soil and Profitability," 2022; Randall, "Turning Former Cropland into Green Grass—and Green Cash," 2021; Fatka, "USDA enhances CRP for climate mitigation," 2021.

240   *Soil sequestration of carbon: Kiss the Ground,* documentary film, 2020; Savory, "How to fight desertification and reverse climate change," 2014; Nordborg, & Röös, "Holistic management—a critical review of Allan Savory's grazing method," 2016; Garnett, *et al., Grazed and confused?* 2017; US Department of Agriculture, "USDA Launches First Phase of Soil Carbon Monitoring Efforts through CRP," 2022.

240   *Recreational value of land:* Hampl, Delmar & Kevin Hampl, Discussion October 28, 2022; Benson, "Buying Recreational Land 12 Benefits (2023) That May Surprise You," accessed 2022; Kansas Department of Wildlife and Parks, "Pheasant," accessed 2022; The National Wild Pheasant Conservation Plan, "Habitat Needs," 2021.

Epilogue

243-244   *2021 prairie fire:* Hampl, Delmar & Kevin Hampl, Discussion October 28, 2022; Kite, "Ball of Rolling Fire and Smoke—Wildfires Rip Through North Central Kansas," 2021; Stafford, "Fires Remain a Concern across Kansas After Strong Wind Storm," 2021.

244-245   *Stone posts:* Hampl, Delmar, Notes to Kent Crawford, October 28, 2022; Hampl, Delmar & Kevin Hampl, Discussion October 28, 2022.

REFERENCES

Abandoned Rails, "The Plainville Branch—Salina to Oakley, KS." Accessed April 20, 2022. https://www.abandonedrails.com/plainville-branch.

Agronomy eUpdates, "Kansas Bankers Association Conservation Awards Program." September 4, 2020. Accessed August 28, 2022. https://eupdate.agronomy.ksu.edu/article_new/kansas-bankers-association-conservation-awards-program-405-7.

Alexander, Kathy. "Territorial Kansas & the Struggle for Statehood." Legends of Kansas, February, 2022. Accessed June 17, 2022. http://legendsofkansas.com/kansas-territory/.

American History Stories and Activities, "Life in a Sod House, Our Story." Smithsonian National Museum of American History. Accessed April 20, 2022. https://amhistory.si.edu/ourstory/activities/sodhouse/more.html.

Anderson, Fred. *Crucible of War: The Seven Years' War and the Fate of Empire in British North America, 1754–1766.* NewYork, Knopf, 2000.

Archives.gov. "Homestead Act of 1862." Accessed March 4, 2022. https://www.archives.gov/files/calendar/genealogy-fair/2014/handouts/session-11-handout-4of5-martinez-land-homestead-act-1862.pdf.

Auto editors of Consumer Guide. "1948-1952 Ford F-Series Trucks." HowStuffWorks. Accessed June 12, 2021. https://auto.howstuffworks.com/1948-1952-ford-fseries-trucks.html.

Bankencyclopedia.com. "The Peoples State Bank of Luray, Kansas in Luray, Kansas (KS)." Banks in Kansas (KS). Accessed April 22, 2022. http://www.bankencyclopedia.com/The-Peoples-State-Bank-of-Luray-Kansas-12430-Luray-Kansas.html.

Batt, Florence. *The History of the Russell County Rural Schools*. Self-published, 1994.

Baxter, Henry H. "The Slip Form Method and Reinforced Concrete Grain Elevators." Accessed November 6, 2022. https://buffaloah.com/h/elev/concrete/conc.html.

Bedford, Laurie. "How Automation Will Transform Farming." *Successful Farming*, November 29, 2017. Accessed January 2, 2023. https://www.agriculture.com/technology/robotics/how-automation-will-transform-farming.

Bellis, Mary. "History of American Agriculture." ThoughtCo, July 3, 2019. Accessed January 20, 2021. https://www.thoughtco.com/history-of-american-agriculture-farm-machinery-4074385.

Bennett, Robert A. "Saving a Town." American Banker Magazine, July 1, 2001. Accessed August 3, 2013. http://www.americanbanker.com/magazine/111_7/-153067-1.html; http://www.americanbanker.com/magazine/111_7/-153067-1.html?pg=2; https://www.americanbanker.com/magazine/111_7/-153067-1.html?pg=3.

Benson, Erika. "Buying Recreational Land: 12 Benefits (2023) That May Surprise You." Gokce Capital. Accessed November 7, 2022. https://gokcecapital.com/buying-recreational-land/.

Blackmar, Frank W. (ed.). *Kansas, A Cyclopedia of State History, Embracing Events, Institutions, Industries, Counties, Cities, Towns, Prominent Persons, Etc.* Standard Publishing Company, Chicago, 1912.

Blue Hills Heritage Foundation. "– 2018 Event Inductee – Luray Fish Fry." Accessed March 15, 2021. https://bhhf2018.wixsite.com/bhhf/2018-luray-fish-fry.

Bluestem Quarry and Stoneworks. "Kansas Limestone History." 2007. Accessed November 21, 2021. http://bluestemstoneworks.com/history.htm.

Bookstore, Pauline, and Clarence Crawford. Discussion May 3, 2008.

Bookstore, Pauline, and Delmar Hampl. Hampl Genealogy Notebook, researched and assembled by Pauline (Crawford) Bookstore and Delmar Hampl, last updated 2008.

Bookstore, Pauline. Crawford Genealogy Notebook researched and assembled by Pauline (Crawford) Bookstore, last updated 2008.

———— Discussion July 30-31, 2010.

———— Discussion October 19, 2010.

———— Discussion July 31, 2011.

———— Discussion, date unknown.

———— Handwritten notes April 7, 2009.

———— Telephone discussion February 21, 2010.

Bounds, Doug. "Kansas: A Leader in Wheat, Grain Sorghum, and Beef Production." *Research and Science*, US Department of Agriculture, National Agricultural Statistics Service, July 29, 2021. Accessed September 14, 2022. https://www.usda.gov/media/blog/2019/07/03/kansas-leader-wheat-grain-sorghum-and-beef-production.

Brainard, Rick. "History of Kansas Agriculture." Kansas State History, Federal Writer's Project. Accessed June 18, 2021. https://kspatriot.org/index.php/articles/48-kansas-agriculture/226-history-of-kansas-agriculture.html.

Bureau of Labor and Industrial Statistics. *Fourth Annual Report of the Bureau of Labor and Industrial Statistics*. Topeka, Kansas, January 1, 1889. Accessed June 15, 2022. https://babel.hathitrust.org/cgi/pt?id=hvd.hl4py0&view=1up&seq=5&skin=2021.

Cart, Kristen. "Johnson-Sampson Construction Company." Our Grandfathers' Grain Elevators. Accessed November 7, 2022. https://ourgrandfathersgrainelevators.com/tag/johnson-sampson-construction-company/.

Carter, P. R., D. R. Hicks, E. S. Oplinger, L. G. Bundy, R. T. Schuler, B. J. Holmes. *Grain Sorghum (Milo), Alternative Field Crops Manual.* University of Wisconsin-Extension and Cooperative Extension, University of Minnesota, 1989. Accessed July 3, 2021. https://www.hort.purdue.edu/newcrop/afcm/sorghum.html.

Classic & Collector Cars, "1919 Ford Model T Touring." Accessed June 21, 2022. https://www.classicandcollectorcars.com/vehicles/206/1919-ford-model-t-touring.

Clow, Albert. "City of Waldo, Kansas." Blue Skyways, Kansas State Library, January 10, 2003. Accessed May 25, 2011. http://skyways.lib.ks.us/towns/Waldo/history.html.

——— "City of Waldo, Kansas," April 17, 2011. Accessed May 22, 2011. http://skyways.lib.ks.us/towns/Waldo/.

Connelley, William E. *A Standard History of Kansas and Kansans, Revised ed.* Lewis Publishing Co., Chicago, 1919. Accessed June 22, 2022. http://www.ksgenweb.org/archives/1919ks/p/pangbuwe.html.

Conrad, Krista. "How Do the Rocky Mountains Influence Climate." World Atlas, 2020. Accessed January 16, 2021. https://www.worldatlas.com/articles/how-do-the-rocky-mountains-influence-climate.html.

Crawford, Clarence. "Building a Post Rock Fence." *Plains Speaking.* Charlotte Crawford, Editor, Post Rock Press, Knoxville, Tennessee, 2021.

——— Century Farm Citation, 2001.

——— Clarence Crawford letters to Gladys Crawford, 1941-1944.

Crawford, Gladys. Chegwidden Genealogy Notebook, researched and assembled by Gladys (Chegwidden) Crawford, last updated 1997.

——— Gladys (Chegwidden) Crawford letters to her sister Helen Hampl, 1941.

——— Gladys (Chegwidden) Crawford notes about her life, 1993.

——— Sechtem Genealogy Notebook, researched and assembled by Gladys (Chegwidden) Crawford, last updated 1997.

Crawford, Harmon. Homestead patent document, July 3, 1890.

REFERENCES

Crawford, R. Kent. *Ruts, Guts, & a Model T Truck: Cruising the West at 15 Miles per Hour*. Post Rock Press, Knoxville, Tennessee, 2021.

Crawford, Roy. Roy Crawford estate tax forms, 1960.

—— Roy Crawford gift tax documents, 1950.

—— Roy Crawford will.

Cutler, William G. *History of the State of Kansas*. "Osborne County, Part 2." A. T. Andreas, Chicago, Illinois, 1883. Transcribed in https://kancoll.org/books/cutler/osborne/osborne-co-p2.html.

—— *History of the State of Kansas*. "Russell County, Part 1.", A. T. Andreas, Chicago, Illinois, 1883. Transcribed in http://www.kancoll.org/books/cutler/russell/russell-co-p1.html.

—— *History of the State of Kansas*. "Russell County, Part 2." A. T. Andreas, Chicago, Illinois, 1883. Transcribed in http://www.kancoll.org/books/cutler/russell/russell-co-p2.html.

—— *History of the State of Kansas*. "Russell County, Part 3." A. T. Andreas, Chicago, Illinois, 1883. Transcribed in http://www.kancoll.org/books/cutler/russell/russell-co-p3.html.

—— *History of the State of Kansas*. "Russell County, Part 5." A. T. Andreas, Chicago, Illinois, 1883. Transcribed in http://www.kancoll.org/books/cutler/russell/russell-co-p5.html.

Czech Friends. "Why they went—Reasons for Czech emigration to America in the 19th century." Accessed January 10, 2021. https://www.czechfriends.org/reasons-for-czech-emigration.

Dalrymple, Dana G. "Changes in Wheat Varieties and Yields in the United States, 1919-1984." Agricultural History, Vol. 62, No. 4, 1988, Agricultural Historical Society. Accessed May 14, 2021. https://pdf.usaid.gov/pdf_docs/pnabb929.pdf.

Dreibus, Tony. "Goodbye, Kansas Wheat?" *Successful Farming*, Meredity Agrimedia, March 26, 2018. Accessed September 15, 2022. https://www.agriculture.com/crops/wheat/goodbye-kansas-wheat.

Dunaway, Wayne Fuller. *The Scotch-Irish of Colonial Pennsylvania*, University of North Carolina Press, 1944. https://collection1.libraries.psu.edu/cdm/ref/collection/digitalbks2/id/18373.

EDinformatics. "Threshing Machine." Accessed April 21, 2022. https://www.edinformatics.com/inventions_inventors/threshing _machine.htm.

Encyclopaedia Britannica. "Czechoslovakia." May 12, 2020. Accessed July 8, 2022. https://www.britannica.com/place/Czechoslovakia.

———— "Kinetoscope." March 3, 2020. Accessed August 18, 2022. https://www.britannica.com/technology/Kinetoscope.

Encyclopedia.com. "Farm Foreclosures." Accessed May 1, 2021. https://www.encyclopedia.com/economics/encyclopedias-almanacs-transcripts- and-maps/farm-foreclosures.

Fandom. "McCormick Deering 15-30." Tractor & Construction Plant Wiki Accessed August 25, 2021. https://tractors.fandom.com/wiki/McCormick-Deering_15-30.

Farmland Information Center. "Kansas Data and Statistics." Accessed May 13, 2021. https://farmlandinfo.org/statistics/kansas-statistics/.

Fatka, Jacqui. "USDA enhances CRP for climate mitigation." Delta Farm Press, April 22, 2021. Accessed November 5, 2022. https://www.farmprogress.com/carbon/usda-enhances-crp-climate-mitigation.

Find-A-Grave "Bertha Maude Reiss Beatty (1914-1982)." Find A Grave Memorial. accessed June 16, 2022. https://www.findagrave.com/memorial/41419803/bertha-maude-beatty.

———— "Daniel James O'Connor (1843-1915)." Find a Grave Memorial. Accessed August 13, 2021. https://www.findagrave.com/memorial/191663733/daniel-james-o'connor.

———— "John A. O'Leary, Jr (1926-1989)." Find a Grave Memorial. Accessed March 12, 2022. https://www.findagrave.com/memorial/83611953/john-a-o'leary.

———— "John A. O'Leary, Sr (1901-1984)." Find A Grave Memorial. Accessed March 12, 2022. https://www.findagrave.com/memorial/83615900/john-a-o'leary.

———— "Lloyd Raymond Beatty (1912-1984)." Find A Grave Memorial. Accessed June 16, 2022. https://www.findagrave.com/memorial/45537036/lloyd-raymond-beatty.

———— "Thomas Anderson Beatty (1830-1909)." Find A Grave Memorial. Accessed August 29, 2021. https://www.findagrave.com/memorial/45536906/thomas-anderson-beatty.

Ganzel, Bill and Claudia Reinhardt. "Postwar Fertilizer Explodes." Farming in the 1940s, Wessels Living History Farm, York, Nebraska. Accessed June 16, 2021. https://livinghistoryfarm.org/farminginthe40s/crops_04.html.

Ganzel, Bill. "Farm Bust of the 1980s." Farming 1970s to Today, Wessels Living History Farm, York, Nebraska, 2009. Accessed September 26, 2021. https://livinghistoryfarm.org/farminginthe70s/money_05.html.

———— "Harvest Technology." Farming 1970s to Today, Wessels Living History Farm, York, Nebraska, 2009 Accessed October 12, 2021. https://livinghistoryfarm.org/farminginthe70s/machines/harvest-technology/.

———— "The Dust Bowl." Farming in the 1930s, Wessels Living History Farm, York, Nebraska, 2003. Accessed June 22, 2022. https://livinghistoryfarm.org/farminginthe30s/water_02.html.

Garnett, Tara, Cécile Godde, *et al. Grazed and confused?* Food Climate Research Network, Oxford Martin Programme on the Future of Food, and Environmental Change Institute, University of Oxford, 2017. Accessed September 10, 2022. https://www.oxfordmartin.ox.ac.uk/downloads/reports/fcrn_gnc_report.pdf.

Giles, Diane. "Remembering the frigid winter of 1885.", *The Kenosha News*, February 5, 2016. https://www.kenoshanews.com/news/archival-revival-remembering-the-frigid-winter-of-1885/article_afaedfd6-c509-5339-a7a3-7b9e0b902d59.html.

Gould, Leslie. "Cemetery Folklore Studied." *The Hays Daily News*, June 28, 1981.

Hampl, Bill, and Amy Hubbard. Oral interview recorded by Gladys and Clarence Crawford, 1980.

Hampl, Delmar, and Kevin Hampl, Discussion October 28, 2022.

Hampl, Delmar. Discussion July 30, 2021.

———— Notes to Kent Crawford, November 25-29, 2010.

———— Notes to Kent Crawford, October 28, 2022.

Hampl, Sarah. Sarah (Griffin) Hampl notes about her life, date unknown. Copy provided by Delmar Hampl on November 9, 2010.

Hansen, M. V. "How a Threshing Machine Works." Farm Collector Newsletter. Accessed June 19, 2021.
https://www.farmcollector.com/equipment/how-a-threshing-machine-works/.

Hergert, Gary, Rex Nielsen, Jim Margheim. " A Historical Overview of Fertilizer Use, parts 1, 2 and 3." University of Nebraska at Lincoln Panhandle Research and Extension Center, March 15, 2015. Accessed June 27, 2021.
https://cropwatch.unl.edu/fertilizer-history-p1;
https://cropwatch.unl.edu/fertilizer-history-p2;
https://cropwatch.unl.edu/fertilizer-history-p3.

Herring, Mary. "Harvest Equipment: A Brief History of the Combine." May 4, 2020. Accessed August 31, 2021.
https://ironsolutions.com/agriculture-equipment-value-guides/a-brief-history-of-the-combine/.

History Colorado. "WPA Privy (1935-1943)." Accessed February 25, 2021.
https://www.historycolorado.org/wpa-privy-1935-1943.

Holt, Kristin. "Victorian Blizzards, Nonstop in the 1880s." March 11, 2017. Accessed August 14, 2021.
http://www.kristinholt.com/archives/10240.

Horses. "How Much Did a Horse Cost in the 1800s." Horses: Everything about Horses. Accessed June 6, 2022.
https://www.skipperwbreeders.com/interesting/how-much-did-a-horse-cost-in-the-1800s.html.

IDTechEX.com. "Agricultural Robots and Drones 2017-2027: Technologies, Markets, and Players." Research Reports, March, 2017. Executive Summary accessed January 2, 2023. https://www.idtechex.com/en/research-report/agricultural-robots-and-drones-2017-2027-technologies-markets-players/525.

Indiana Historical Society. "Northwest Territory Papers and Documents, 1721-1802 (Bulk 1780-1801)." Collection # M 0367 OMB 0042, Manuscript and Visual Collections Department William Henry Smith Memorial Library, Indiana Historical Society, 450 West Ohio Street, Indianapolis, IN. Accessed March 4, 2022 . https://indianahistory.org/wp-content/uploads/northwest-territory-collection.pdf.

Ise, John. *Sod and Stubble,* Unabridged and Annotated Edition with additional material by Von Rothenberger, University Press of Kansas, Lawrence, Kansas, 1996. This is an annotated version of the original John Ise book, *Sod and Stubble*, published in 1936.

Isern, Thomas D. "Adoption of the Combine on the Northern Plains." South Dakota State Historical Society, 1980. Accessed August 23, 2021. https://www.sdhspress.com/journal/south-dakota-history-10-2/adoption-of-the-combine-on-the-northern-plains/vol-10-no-2-adoption-of-the- combine-on-the-northern-plains.pdf.

Juracek, Kyle E., Charles A. Perry, James E. Putnam. "The 1951 Floods in Kansas Revisited." US Geological Survey, May 2001. Accessed March 12, 2021. https://pubs.usgs.gov/fs/2001/0041/report.pdf.

Kansapedia. "Blizzard of 1886." Kansas State Historical Society, June 2011. Accessed January 14, 2021. https://www.kshs.org/kansapedia/blizzard-of-1886/11982.

———"Opera Houses." Kansas Historical Society. Accessed February 29, 2021. https://www.kshs.org/kansapedia/opera-houses/14232.

Kansas Ag Growth. "Wheat." 2019. Accessed September 14, 2022. https://agriculture.ks.gov/docs/default-source/ag-growth-summit/january-2018-documents/wheat-sector.pdf.

Kansas Department of Agriculture. "Kansas Farm Facts." 2019. Accessed May 1, 2021. https://agriculture.ks.gov/docs/default-source/ag-marketing/kansas-farm-facts-2019.pdf?sfvrsn=32b689c1_4.

Kansas Department of Wildlife and Parks. "Pheasant." Accessed November 8, 2022. https://ksoutdoors.com/Hunting/Upland-Birds/Pheasant.

Kansas Farm Bureau. "Do You Live on a Century Farm?" Accessed March 19, 2021. https://www.kfb.org/Get-Involved/Contests-Recognition/Century-Farm-Program.

Kansas High School Football History. "Lucas-Luray Cougars." Accessed June 29, 2022. http://www.kansasfootballhistory.com/teams.cfm?school=lucas-luray.

Kansas Historical Society. "Jackrabbit Drives." April, 2015. Accessed June 6, 2022. http://www.kshs.org/kansapedia/jackrabbit-drives/12097.

——— "Kansas Territory." Kansapedia, April, 2010. Accessed June 17, 2022. http://kshs.org/kansapedia/kansas-territory/14701.

Kansas State College. *The Royal Purple for 1936*. Manhattan, Kansas.

Kansas State University. *Grain Sorghum Production Handbook*. Kansas State University Agricultural Experiment Station and Cooperative Extension Service, Manhattan, Kansas, C-687 1998. https://bookstore.ksre.ksu.edu/pubs/C687.pdf.

——— "K-State Army ROTC History." Military Science (Army ROTC), Manhattan, Kansas, May 25, 2018. Accessed September 20, 2021. https://www.k-state.edu/military-science/.

——— "What is K-State Research and Extension?" Wild West District, K-State Research and Extension. Accessed May 1, 2021. https://wildwest.k-state.edu/about/.

Kansas. "1885 Kansas Agricultural Census." 1885 Census, State of Kansas, Schedule 2—Agriculture, March, 1885. Kansas Historical Society.

——— "1895 Kansas Agricultural Census." 1895 Census, State of Kansas, Schedule 2—Agriculture, March, 1895. Kansas Historical Society.

Kantor, Sylvia. "Conserving Soil and Water in Dryland Wheat Region." No-Till Farmer, January 23, 2015. Accessed April 21, 2022. https://www.no-tillfarmer.com/articles/4335-study-conserving-soil-and-water-in-dryland-wheat-region.

Kaufman, David A. "209th AAA Automatic Weapons Battalion (Self-Propelled)." 2001. Accessed August 31, 2014.
http://www.skylighters.org/aaapatches/209.html.

KCKPL. "Historic School Buildings of Kansas City, Kansas." Online Programs @ KCKPL. Accessed January 10, 2022.
https://kckplprograms.org/2021/07/29/historic-school-buildings-of-kansas-city-kansas/.

King, Newton. E-mail to Kent Crawford, July 28, 2012.

Kiss the Ground. Documentary film; executive producers: Julian Lennon and Gisele Bündchen, narrated by Woody Harrelson, released on Netflix in 2020. https://kissthegroundmovie.com/.

Kite, Allison. "'Ball of Rolling Fire and Smoke'–Wildfires Rip Through North Central Kansas." Kansas Reflector, December 20, 2021. Accessed Nov. 8, 2022.
https://kansasreflector.com/2021/12/20/ball-of-rolling-fire-and-smoke-wildfires-rip-through-north-central-kansas/.

K-State Alumni Association. "K-State History." Accessed May 1, 2021.
https://www.k-state.com/about/kstatehistory.php.

K-State Research and Extension. "Chisholm Trail District." Accessed April 22, 2022. https://www.chisholmtrail.k-state.edu/.

——— *Wheat Production Handbook*, C-529, 1997.
https://bookstore.ksre.ksu.edu/pubs/C529.pdf.

Kuhn North America. "History in North America." Accessed June 24, 2021. https://www.kuhn-usa.com/about-kuhn/about-us/kuhn-north-america/history-north-america.

Kysilka, Karel. "Emigration to the USA from the Policka region in 1850 – 1890." Genealogy Seminar of the Czech Heritage Society of Texas, Hillsboro, TX, July 31, 1999. Accessed January 10, 2021.
http://www.fortunecity.com/victorian/durer/23/emigration/emigration.htm.

Lee, Robert. "The True Cost of the Louisiana Purchase." *Slate*, March 1, 2017 Accessed March 4, 2021.
http://www.slate.com/articles/news_and_politics/history/2017/03/how_much_did_the_louisiana_purchase_actually_cost.html.

Legends of Kansas. "Black Friday Flood, 1951." Accessed November 26, 2022. https://legendsofkansas.com/black-friday-flood/.

———— "Quindaro, Kansas–a Free-State Black Town." Accessed January 10, 2022. https://legendsofkansas.com/quindaro-kansas/.

Library of Congress. "Map of Nebraska Showing the Union Pacific Railroad Land Grant." 1880. Accessed June 17, 2022. http://loc.gov/resource/g4191p.rr005930/.

Living History Farms. "Grain Harvest and Threshing Time.", July 28, 2015. Accessed June 19, 2022. http://lhf.org/2015/07/grain-harvest-and-threshing-time/.

Lotterman, Edward. "Farm Bills and Farmers: The effects of subsidies over time." Federal Reserve Bank of Minneapolis, December 1, 1996. Accessed July 15, 2021. https://www.minneapolisfed.org/article/1996/farm-bills-and-farmers-the-effects-of-subsidies-over-time.

Luder, John. "The Golden Prairies of the Middle West." Typed notes probably published in a Waldo school newspaper.

Luder, Laura. Letters to her niece Pauline (Crawford) Bookstore, 1969.

Luray Herald. "Moving Picture Show." December 19, 1902, p. 1.

Lutz, Arlen, and Jack Aschwege. "EC59-806 Custom Harvesting Rates Paid in Nebraska 1958." University of Nebraska–Lincoln, 1959. Accessed March3, 2021. https://digitalcommons.unl.edu/cgi/viewcontent.cgi?article=4456&context=extensionhist.

Malin, James C. "Dust Storms: Part Two, 1861-1880.", Kansas Historical Society, August 1946, Vol. 14, No. 3, pp. 265 -296. Accessed January 15, 2021. https://www.kshs.org/p/kansas-historical-quarterly-dust-storms-part-two-1861-1880/13031.

———— "Dust Storms: Part Three, 1881-1900." Kansas Historical Society, November 1946, Vol. 14, No. 4, pp. 391 -413. Accessed January15, 2021. https://www.kshs.org/p/dust-storms-part-three-1881-1900/13039.

Martin, Fernando M. "A Short History of Prices, Inflation since the Founding of the U.S." Federal Reserve Bank of Saint Louis, July 25, 2017. Accessed April 25, 2021.

https://www.stlouisfed.org/publications/regional-economist/second-quarter-2017/a-short-history-of-prices-inflation-since-founding-of-us.

McCullough, David. *The Pioneers*. Simon & Schuster, New York, 2019.

McFadden, Jonathan, "Drought-Tolerant Corn in the United States: Research, Commercialization, and Related Crop Production Practices." US Department of Agriculture, Economic Research Service, *Amber Waves Magazine*, March 13, 2019. Accessed September 17, 2022. https://www.ers.usda.gov/amber-waves/2019/march/drought-tolerant-corn-in-the-united-states-research-commercialization-and-related-crop-production-practices/

McMahon, Tim. "Historical Consumer Price Index (CPI-U) Data." Accessed April 13, 2021. https://inflationdata.com/Inflation/Consumer_Price_Index/HistoricalCPI.aspx?reloaded=true.

McManus, Leslie C. "Steam Threshing Rings." Farm Collector. Accessed August 23, 2021. https://www.farmcollector.com/steam-engines/steam-threshing-rings/.

Military History Matters. "The Franco-Prussian War." November 12, 2020. Accessed June 23, 2022. https://www.military-history.org/cover-feature/the-franco-prussian-war.htm.

Miller, Hunter (ed.). "British-American Diplomacy: The Paris Peace Treaty of September 30, 1783." The Avalon Project at Yale Law School. Accessed June 13, 2022. http://avalon.law.yale.edu/18th_century/paris.asp.

Miller, Kaitlin. "The Most Devastating Winter Storms in US History.". The Active Times, February 6, 2020. Accessed August 19, 2021. https://www.theactivetimes.com/home/featured/worst-winter-storms-us-history.

Miller, Lynn R. "A Short History of the Horse-Drawn Mower." Small Farmer's Journal. Accessed February 5, 2022. https://smallfarmersjournal.com/a-short-history-of-the-horse-drawn-mower/.

Miner, Craig. *West of Wichita*. University Press of Kansas, Lawrence, Kansas,1986, Chapter 1.

Mottech. "Why Automating Agriculture is the Future of Farming." Motorola, April 3, 2022. Accessed January 2, 2023. https://mottech.com/news/automating-agriculture-the-future-of-farming/.

My Czech Roots, "Emigration—Basic History." Accessed January 10, 2021. https://www.myczechroots.com/records/emigration.

Nace, R. L., and E. J. Pluhowski. "Drought of the 1950's with Special Reference to the Midcontinent." Geological Survey Water-Supply Paper 1804, US Government Printing Office, Washington, 1965. Accessed March 12, 2021. https://pubs.usgs.gov/wsp/1804/report.pdf.

National Agricultural Law Center, "Conservation Programs—An Overview." Accessed January 31, 2021. https://nationalaglawcenter.org/overview/conservation-programs/.

National Drought Mitigation Center. "The Dust Bowl, University of Nebraska – Lincoln. Accessed April 30, 2021. https://drought.unl.edu/dustbowl/Home.aspx.

National Museum of American History. "Chromatolithograph entitled 'Custer's Last Fight'." Accessed January 7, 2023. https://americanhistory.si.edu/collections/search/object/nmah_3 26129.

National Weather Service. "Hemispheric Prevailing Winds." Flight Environment, National Weather Service, National Oceanic and Atmospheric Administration. Accessed June 18, 2022. http://weather.gov/source/zhu/ZHU_Training_Page/winds/Wx_Terms/Flight_Environment.htm.

———— "Origin of Wind." National Weather Service-Jet Stream, National Oceanic and Atmospheric Administration. Accessed June 18, 2022. http://weather.gov/jetstream/wind.

———— "60th Anniversary of the March 23-25, 1957 Blizzard." National Oceanic and Atmospheric Administration, Dodge City, KS

Weather Forecast Office, 2021. Accessed March 12. 2021. https://www.weather.gov/ddc/1957Blizzard.

Nickel, Raylene. "Converting Marginal Cropland to Perennials Builds Soil and Profitability." *Successful Farming*, Meridith Agrimedia, July 23, 2022. Accessed September 15, 2022. https://www.agriculture.com/crops/conservation/converting-marginal-cropland-to-perennials-builds-soil-and-profitability.

Nordborg, Maria and Elin Röös. "Holistic management—a critical review of Allan Savory's grazing method." ResearchGate, June, 2016. Accessed September 16, 2022. https://www.researchgate.net/publication/309589057_Holistic_m anagement_-_a_critical_review_of_Allan_Savory's_grazing_method.

Northampton County Historical and Genealogical Society. *The Scotch-Irish of Northampton County, Pennsylvania*. Easton, Pennsylvania, 1926.

NorthWest Publishing Co. *Plat Book of Russell County Kansas*. Minneapolis, Minnesota, 1901.

Nothern, Joan. "Cattle Trails, Rodeos & Livestock Stories—Cloud County." Solomon Valley Highway 24 Heritage Alliance. Accessed April 20, 2022. https://www.hwy24.org/uploads/2/6/1/8/26189167/14_cattletrails _rodeos_livestock_stories.pdf.

Osborne County. "The Creation and Organization of Osborne County Kansas." Osborne County: News and Information: History. Accessed October 2, 2021. http://www.osbornecounty.org/news_and_information/history/.

Oswald, John. Communication November 15, 2020.

Ozatwar.com, "94th Coast Artillery (AA) Regiment 40th Anti-Aircraft Brigade in Australia During WW2." Accessed August 29, 2014. http://www.ozatwar.com/usarmy/94thcar.htm.

Paulsen, Gary M. "Hard White Winter Wheat for Kansas." Kansas State University Agricultural Experiment Station and Cooperative Extension Service, Manhattan, Kansas, SRL-120, March, 1998.

Accessed April 30, 2021. https://www.ksre.k-state.edu/historicpublications/pubs/SRL120.pdf

Pearson, Calvin H., and Amaya Atucha. "Agricultural Experiment Stations and Branch Stations in the United States." *Journal of Natural Resources and Life Sciences Education*, December 18, 2015. Accessed April 30, 2021. https://acsess.onlinelibrary.wiley.com/doi/abs/10.4195/nse2013.10.0032.

Penner, Mil. *Section 27: A Century on a Family Farm*. University Press of Kansas, Lawrence, Kansas, 2002, Chapter 11.

Plainville Times. "Natoma USD-391 & Paradise-Waldo USD-399 School Patrons Favor Joining Districts." February 14, 1974, p. 6.

Point to Point Surveyors. "The Louisiana Purchase Survey." July 6, 2016. Accessed December 25, 2020. https://www.pointtopointsurvey.com/2016/07/louisiana-purchase-survey/.

Powell, G. Morgan, and Kent McVay. *Terrace Maintenance*. K-State Research and Extension, Departments of Agronomy and Agricultural Engineering, C709, July, 2004. Accessed March 3, 2021. https://www.coffey.k-state.edu/crops-livestock/crops/conservation/Terrace%20Maintenance.pdf.

Randall, Brianna. "Turning Former Cropland into Green Grass—and Green Cash." US Department of Agriculture, September 15, 2021. Accessed September 15, 2022. https://www.farmers.gov/blog/turning-former-cropland-into-green-grass-and-green-cash.

Rasmussen, Wayne D., Gladys L. Baker, James S. Ward. *A Short History of Agricultural Adjustment, 1933-75*. National Economics Analysis Division, Economic Research Service, US Department of Agriculture, Washington, D.C., 1976. Information Bulletin No. 391. Accessed January 29, 2021. https://naldc.nal.usda.gov/download/CAT87210025/PDF.

Reinhardt, Claudia. "Harvesting Wheat." Farming in the 1920s, Wessels Living History Farm, York, Nebraska. Accessed August 4, 2021.

https://livinghistoryfarm.org/farminginthe20s/intro/machine/har vesting-wheat/.

Reitz, L. P., and H. H. Laude. "Comanche and Pawnee: New Varieties of Hard Red Winter Wheat for Kansas.", Agricultural Experiment Station, Kansas State College of Agriculture and Applied Science, Manhattan, Kansas, July, 1943. Accessed May 3, 2021. https://www.ksre.k-state.edu/historicpublications/pubs/SB319.PDF.

Revivaler. "Aultman and Taylor 30-60." Accessed August 26, 2021. https://revivaler.com/aultman-and-taylor-30-60/.

Ridgway, Mary. "Remembering Union Pacific's Plainville Branch." Railroads in Graham County: 1888—1998, Graham County Historical Society, 1998. Accessed May 25, 2011. http://www.grahamhistorical.ruraltel.net/trains/trains.html.

Robinson, Tammy. "Winter Survival Skills That Kept The Pioneers Alive." Off The Grid News. Accessed April 29, 2022. https://www.offthegridnews.com/lost-ways-found/winter-survival-skills-that-kept-the-pioneers-alive/.

Root, George A. "Ferries in Kansas, Part VII, Saline River." The Kansas Historical Quarterly, Volume IV, No. 2, May 1935, Kansas Historical Society. Accessed May 6, 2022. https://www.kshs.org/p/kansas-historical-quarterly-ferries-in-kansas-part-vii-saline-river/12654.

Rural Electrification Administration. *Rural Lines, USA: the story of the Rural Electrification Administration's first 25 years, 1935-1960.* United States Rural Electrification Administration, 1960. Accessed April 21, 2022. https://ia803200.us.archive.org/27/items/rurallinesusasto811unit_0/rurallinesusasto811unit_0.pdf.

——— "Rural Lines—USA, The Story of Cooperative Rural Electrification." Miscellaneous Publication Number 811, Rural Electrification Administration, US Department of Agriculture, 1981. Accessed March 7, 2022. https://ageconsearch.umn.edu/record/319871.

Russell County Economic Development & Convention and Visitors Bureau. "56th Annual Luray Friendship Day." Accessed September 6, 2022. https://www.russellcountyks.org/events-1/56th-annual-luray-friendship-day.

Russell County Historical Society. Notes about Waldo history.

Russell County Recorder of Deeds office. Land records. Accessed frequently, 2013-2018.

Russell Daily News. "Goodyear Award presented to Clarence Crawford." Russell, Kansas, February 11, 1976.

———— "J. W. Van Scoyoc Third to File Claim in Russell County." Russell, Kansas, June 3, 1961.

———— "Luray Man New Member of State Education Board." Autumn 1959.

Russell Record. "House of Seven Gables." September 13, 1977.

Savory, Allan. "How to fight desertification and reverse climate change." Ted Talk, September 17, 2014. Accessed September 22, 2022. http://www.ted.com/talks/allan_savory_how_to_green_the_world_s_deserts_and_reverse_climate_change.

Schaller, Natalie. Thomas Griesser, Andreas Marc Fischer, Alexander Stickler, Stefan Brönnimann. "Climate effects of the 1883 Krakatoa eruption: Historical and present perspectives." Vierteljahrsschrift der Naturforschenden Gesellschaft in Zürich, January, 2009 Accessed August 14, 2021. https://www.researchgate.net/publication/255700466_Climate_effects_of_the_1883_Krakatoa_eruption_Historical_and_present_perspectives.

Schmitt, Erin. "The Shape of Water Towers—An Engineering History." Accessed Nov. 7, 2022. https://www.tpomag.com/online_exclusives/2018/07/the-shape-of-water-towers-an-engineering-history.

Schultz Family. *Luray High School 1914-1977, Alumni Memory Book*. May, 2012. Copy provided by John Oswald.

Sechtem, Caspar. Caspar Sechtem army discharge papers, Caspar Sechtem original documents, 1877.

Shipley, Samuel. "list of U.S. states' dates of admission to the union." Encyclopedia Britannica, Inc., February 11, 2020. Accessed March 4, 2022. http://britannica.com/topic/list-of-U-S-states-by dates-of-admission-to-the-union-2130026.

Snapp, Sieg, and Vicki Marone. "Perennial Wheat." Michigan State University Extension Bulletin E3208, February, 2014. Accessed September 13, 2022.
https://www.canr.msu.edu/resources/perennial_wheat

Soil Conservation Service. "Conservation Plan, Clarence R. Crawford Operator, Roy A. Crawford Owner, Agreement No. 57, US Department of Agriculture Soil Conservation Service, April 8, 1948."

———— "Soil and Water Conservation Plan, Clarence R. Crawford Operator, Pauline Bookstore Owner, Russell County Soil Conservation District, US Department of Agriculture Soil Conservation Service, February 15, 1965."

———— "Soil and Water Conservation Plan, Clarence R. Crawford Trust #1 Operator, Pauline Bookstore Owner, Russell County Soil Conservation District, US Department of Agriculture Soil Conservation Service, July 6, 1988."

Soucy, D. L. "Dugouts and Sod Houses—How to Build Them." Surviving the Times, October 6, 2010. Reprint of article by A. F. Wallace, Hunter-Trapper-Trader magazine, July, 1911. Accessed April 20, 2022. https://dlsoucy.wordpress.com/2010/06/10/dugouts-and-sod-houses-how-to-build-them/.

South Dakota State University. "One-Way Disc Plow—Adjustment and Operation." SDSU Extension Fact Sheets 544, 1977. Accessed April 21, 2022.
https://openprairie.sdstate.edu/cgi/viewcontent.cgi?article=1544&context=extension_fact.

Stafford, Margaret. "Fires Remain a Concern Across Kansas After Strong Wind Storm." Associated Press, December 16, 2021. Accessed Nov. 8, 2022. https://www.usnews.com/news/best-states/kansas/articles/2021-12-16/fires-remain-a-concern-across-kansas-after-strong-wind-storm.

Steel in the Field. "Stubble Mulch Blade Plow." Sustainable Agriculture Research and Education, 2001. Accessed June 16, 2021. https://www.sare.org/publications/steel-in-the-field/dryland-crop-tools/stubble-mulch-blade-plow/.

St. John News. "Bookstore-Crawford Wedding." St. John, Kansas, December 1945.

Suchy, Daniel R. "The Public land Survey System in Kansas." Public Information Circular (PIC) 20, January, 2002. Accessed December 25, 2020.
http://www.kgs.ku.edu/Publications/pic20/pic20_2.html.

The History Museum. "The Oliver Corporation." South Bend, Indiana. Accessed August 31, 2021.
https://www.historymuseumsb.org/the-oliver-corporation/.

The Land Institute. "Kernza Grain." Salina, Kansas, 2022. Accessed September 13, 2022.   https://landinstitute.org/our-work/perennial-crops/kernza/.

The Living New Deal. "Agricultural Adjustment Act (1933, reauthorized 1938)." November 18, 2016. Accessed January 29, 2021. https://livingnewdeal.org/glossary/agricultural-adjustment-act-1933-re-authorized-1938-2/.

The Museum of Yesterday. "Antique Wireless Equipment Collection—Telephone." Chesterfield, Virginia. Accessed April 21, 2022. http://www.museumofyesterday.org/museum/page3.htm.

The National Wild Pheasant Conservation Plan & Associated Partnerships. "Habitat Needs." 2021. Accessed November 8, 2022. https://nationalpheasantplan.org/habitat-needs/.

Thompson, David. "My Massey combines from 1948 to 1991." Grainews, Machinery & Shop, February, 2020. Accessed June 12, 2021. https://www.grainews.ca/machinery-shop/my-massey-combines-from-1948-to-1991/.

Topeka Capital Journal. "Schools ask to consolidate." cjonline, January 7, 2010. Accessed June 29, 2022.
https://www.cjonline.com/story/news/local/2010/01/07/schools-ask-to-consolidate/16510498007.

Tractor Data. "Aultman & Taylor 30-60." Accessed August 26, 2021. http://www.tractordata.com/farm-tractors/005/4/5/5455-aultman-&-taylor-30-60.html.

Trainorders. "UP: Remembering the Plainville Branch." January 31, 2019. Accessed February 21, 2021. https://www.trainorders.com/discussion/read.php?11,4722607.

Trimble, Marshall. "What was the Fare for Railroads and Stagecoaches?" True West Magazine, June 25, 2018. Accessed August 12, 2021. https://truewestmagazine.com/fare-railroads-stagecoaches/.

Tripp, O. W. "Do You Remember when Waldo Had the Following? containing later additions by Freddie Schilling." 1970, Russell County Historical Society.

UMB Bank. Letter to Roy Kent Crawford, June 20, 2011.

Union of Concerned Scientists. "Climate Change and Agriculture—A Perfect Storm in Farm Country." March 20, 2019. Accessed October 16, 2021. https://ucsusa.org/resources/climate-change-and-agriculture.

United States Census Bureau. "US Census Data." Accessed multiple times 2019-2022. https://www2.census.gov/library/publications/decennial/.

United States Environmental Protection Agency. "What Climate Change Means for Kansas." EPA-430-F-16-018, 2016. Accessed October 16, 2021. https://archive.epa.gov/epa/production/files/2016-08/documents/climate-change-ks.pdf.

University of Minnesota. "The One-Way Disc Tiller." Agricultural Engineering News Letter, Agricultural Extension Division, July 15, 1938. Accessed April 21, 2022. https://conservancy.umn.edu/bitstream/handle/11299/177371/mn2000-aenl-no076.pdf;sequence=1.

University of Missouri Libraries. "Price and Wages by Decade." Accessed August 16, 2021. https://libraryguides.missouri.edu/c.php?g=28284&p=174161.

US Bank Locations. "The Farmers Bank of Osborne." Accessed November 19, 2022. https://www.usbanklocations.com/the-farmers-bank-of-osborne-4748.shtml.

———— "The Peoples State Bank of Luray, Kansas." Accessed April 22, 2022. https://www.usbanklocations.com/the-peoples-state-bank-of-luray-kansas-12430.shtml.

US Department of Agriculture. "Commodity Costs and Returns.", Economic Research Service. 2021. Accessed January 15, 2022. https://www.ers.usda.gov/data-products/commodity-costs-and-returns/commodity-costs-and-returns/#Historical%20Costs%20and%20Returns:%20Wheat.

———— "Conservation Reserve Program." Farm Service Agency, 2022. Accessed September 15, 2022. https://www.fsa.usda.gov/programs-and-services/conservation-programs/conservation-reserve-program/index.

———— *Engineering Field Handbook*, "Chapter 8, Terraces." Part 650, Natural Resources Conservation Service, 2011. Accessed May 15, 2022. https://directives.sc.egov.usda.gov/OpenNonWebContent.aspx?content=31181.wba.

———— "ERS Charts of Note." Economic Research Service. 2021. Monday, November 17, 2014 chart and others. Accessed October 15, 2021. https://www.ers.usda.gov/data-products/charts-of-note/charts-of-note/?topicId=14869.

———— "Fertilizer Use and Price." Economic Research Service, 2019. Accessed October 14, 2021. https://www.ers.usda.gov/data-products/fertilizer-use-and-price.aspx.

———— "Grain Drills and Planters.", Technical Note No: TX-PM-16-03, Natural Resources Conservation Service, June, 2016. Accessed June 17, 2021   . https://www.nrcs.usda.gov/Internet/FSE_PLANTMATERIALS/publications/etpmctn12915.pdf.

———— *Kansas Wheat History*. National Agricultural Statistics Service News Release, October, 2017. Accessed March 3, 2021. https://www.nass.usda.gov/Statistics_by_State/Kansas/Publications/Cooperative_Projects/KS-wheat-history19.pdf.

REFERENCES

———— "The Home Demonstration Agent." AIB 38, July, 1951. Accessed June 28, 2022.
https://naldc.nal.usda.gov/download/CAT87791369/PDF.

———— "USDA Launches First Phase of Soil Carbon Monitoring Efforts through CRP." USDA Farm Service Agency, Kansas State Office, November State Office Newsletter, e-mail November 2, 2022. Accessed November 4, 2022.

US Global Change Research Program. *Fourth National Climate Assessment, Volume II.* US Government Publishing Office, Washington, D.C., 2018. Accessed September 23, 2020.
https://nca2018.globalchange.gov.

Vogan, Mark. "A LOOK BACK Greatest of Arctic Outbreaks: 1899." Weather.com, December 27, 2015. Accessed June 20, 2022.
http://www.markvoganweather.com/2015/12/27/a-look-back-greatest-of-arctic-outbreaks-1899/.

Waymarking. "History of Rocky Ford." Rocky Ford, CO, February 20, 2011. Accessed August 31, 2021.
https://www.waymarking.com/waymarks/WMARVZ_History_of_Rocky_Ford_Rocky_Ford_CO.

Webster, Ian. "Inflation Calculator." US Official Inflation Data, Alioth Finance, April 13, 2021. Accessed April 25, 2021.
https://www.officialdata.org/.

Weiser, Kathy. "Kansas Pacific Railway." Legends of America, December, 2020. Accessed March 5, 2022.
https://www.legendsofamerica.com/kansas-pacific-railroad/.

Wendel, Charles H. *Encyclopedia of American Farm Implements and Antiques.* Krause Publications, Iola, Wisconsin, 2004, p. 186. Accessed June 10, 2011.
http://books.google.com/books?id=dJlJAM_hJD0C&pg=PA186&lpg=PA186&dq=McSherry+grain+drill&source=bl&ots=tfdR-pZdoo&sig=TgpCDBabP-sy6o9Q-PDBvV8ROI8&hl=en&ei=mGvyTZSuI9DqgQfoo8XUCw&sa=X&oi=book_result&ct=result&resnum=2&sqi=2&ved=0CDEQ6AEwAQ#v=onepage&q=McSherry%20grain%20drill&f=false.

White, C. Albert. *A History of the Rectangular Survey System*. United States Department of the Interior, Bureau of Land Management, Washington, D.C. 20402, 1983. https://ia801909.us.archive.org/34/items/historyofrectangwhit/historyofrectangwhit.pdf.

White, William J. "Economic History of Tractors in the United States." Research Triangle Institute, Economic History Association, 2008. Accessed July 6, 2021. https://eh.net/encyclopedia/economic-history-of-tractors-in-the-united-states/.

Woods, Warren Chip. *County Road Laws of Kansas*. 2008, revised 2011. Accessed January 14, 2021. https://www.ksls.com/Resources/Documents/County%20Road%20Laws%20of%20Kansas.pdf.

World War II Troop Ships. "Queen Mary – Specific Crossing Information–1942." 2007. Accessed September 3, 2017. http://ww2troopships.com/ships/q/queenmary/crossings1942.htm.

Wrangle "North American Mixed Grass Prairie." 2022. Accessed December 16, 2022. https://www.wrangle.org/ecotype/north-american-mixed-grass-prairie.

Yesterday's Tractor Co. "Antique Tractor Resources - Original Tractor Prices: Case." Accessed June 12, 2021. https://www.yesterdaystractors.com/atrp/list/case.htm.

Your Local Farmer. "The Moldboard Plow: history, uses, and definitions." Farm Machinery, 2020. Accessed March 5, 2022. https://urlocalfarmer.com/the-moldboard-plow-history-uses-and-definitions/.

Zhang, Guorong. "Wheat Breeding." Agricultural Research Center—Hays, Kansas State Agricultural Research Center. Accessed April 30, 2021. https://www.hays.k-state.edu/programs/wheat/index.html.

Media Credits

All figures in this book are from Crawford family photos, documents, or hand-drawn maps, except for the following:

3. Based on a public domain map published by the US Geologic Survey. https://www.usgs.gov/media/images/sixth-principal-meridian.

5. Plow in Russell County Historical Society Museum. Photo by Charlotte Crawford

6. From the New York Public Library. Public domain: The Miriam and Ira D. Wallach Division of Art, Prints and Photographs: Photography Collection, The New York Public Library. "Family in front of a sod house." New York Public Library Digital Collections. Accessed May 13, 2022. https://digitalcollections.nypl.org/items/510d47e0-6340-a3d9-e040-e00a18064a99.

7. Crawford farm photograph taken in 1919, modified to represent the earlier configuration of the stone house as described by Laura (Crawford) Luder.

11. Photo obtained from Eldon Hampl, who lived in an apartment in this building, and operated an electrical appliance sales and repair shop there. (Photo provided by John Oswald.)

12. Photo obtained from Eldon Hampl. (Photo provided by John Oswald.)

42. Disking/Plowing a Field, University of Guelf Library Digital Collections, public domain. https://images.ourontario.ca/Partners/uguelphdc/UoG250485f.jpg.

55. Greg Ptashny, public domain. https://www.publicdomainpictures.net/en/view-image.php?image=43003&picture=&jazyk=JP

56. Luray School yearbook, The Spotlight, 1959 edition, Editor, Lauda Fallis.

62. *Russell Daily News*, Russell, Kansas, February 11, 1976.

66. Photo courtesy of Kevin Hampl.

67. Photo courtesy of Kevin Hampl.

About the Author

R. Kent Crawford grew up in the house built by his great grandmother on the central Kansas farm that his great grandparents homesteaded in the late 1800s. After attending public school in the nearby town of Luray, he obtained a bachelor's degree in physics from Kansas State University and a PhD in physics from Princeton University.

He spent his professional career primarily at Argonne National Laboratory and at Oak Ridge National Laboratory, authoring numerous technical articles and book chapters. He is a Fellow of the American Physical Society. Since his retirement he has spent his spare time researching and writing articles about various aspects of his family's history. His first non-technical book, *Ruts, Guts, and a Model T Truck*, was published in 2021.